Self-Assessment Color Review of
Equine Internal Medicine

Tim S. Mair
BVSc, PhD, MRCVS
Bell Equine Veterinary Clinic, UK

Thomas J. Divers
DVM, Diplomate ACVIM, Diplomate ACVECC
Cornell University, USA

Iowa State University Press/Ames

First published in the United States of America in 1997 by
Iowa State University Press, 2121 South State Avenue, Ames, Iowa 50014-8300
ISBN 1-8138-2864-3
Library of Congress Cataloging-in-Publication Data applied for

Copyright © 1997 Manson Publishing Ltd

All rights reserved. No part of this publication may be reproduced, stored in a retrieval system or transmitted in any form or by any means without the written permission of the copyright holder or in accordance with the provisions of the Copyright Act 1956 (as amended), or under the terms of any licence permitting limited copying issued by the Copyright Licensing Agency, 33–34 Alfred Place, London WC1E 7DP.

Any person who does any unauthorised act in relation to this publication may be liable to criminal prosecution and civil claims for damages.

For full details of all Manson Publishing Ltd titles please write to:
Manson Publishing Ltd, 73 Corringham Road, London NW11 7DL, UK.

Design and layout: EDI
Text editing: Peter Beynon
Colour reproduction: Tenon & Polert Colour Scanning Ltd, Hong Kong
Printed by: Grafos SA, Barcelona, Spain

Contributors

Douglas Byars, DVM,
Diplomate ACVIM,
Hagyard–Davidson–McGee Associates,
Director, Equine Internal Medicine Equine Hospital,
Lexington,
USA

Noah D. Cohen, VMD, MPH, PhD,
Diplomate ACVIM,
Assistant Professor of Equine Medicine,
Department of Large Animal Medicine & Surgery,
College of Veterinary Medicine,
Texas A & M University,
USA

Chrysann Collatos, VMD, PhD,
Diplomate ACVIM,
High Desert Veterinary Service,
Reno,
USA

Thomas J. Divers, DVM,
Diplomate ACVIM,
Diplomate ACVECC,
New York State College of Veterinary Medicine,
Cornell University,
Ithaca,
USA

John M. King, DVM, PhD,
NewYork State College of Veterinary Medicine,
Cornell University,
Ithaca,
USA

Sandy Love, BVMS, PhD, MRCVS,
Professor of Equine Clinical Studies,
Division of Equine Clinical Studies,
University of Glasgow Veterinary School,
Scotland

Tim Mair, BVSc, PhD, MRCVS,
Bell Equine Veterinary Clinic,
Maidstone,
England

Celia M. Marr, BVMS, MVM, PhD, MRCVS,
The Royal Veterinary College,
University of London,
England

William H. Miller, Jr, VMD,
Diplomate ACVD,
New York State College of Veterinary Medicine,
Cornell University,
Ithaca,
USA

Elspeth M. Milne, BVM&S, PhD, MRCVS,
Veterinary Investigation Officer,
SAC Veterinary Services,
Dumfries,
Scotland

Christopher J. Proudman, MA, VetMB, PhD, Cert EO, FRCVS,
Division of Equine Studies,
Department of Clinical Veterinary Medicine,
University of Liverpool,
England

William C. Rebhun, DVM,
Diplomate ACVO,
Diplomate ACVIM,
Professor of Ophthalmology and Large Animal Medicine,
New York State College of Veterinary Medicine,
Cornell University,
Ithaca,
USA

Johanna M. Reimer, VMD,
Diplomate ACVIM, Internal Medicine & Cardiology,
Rood & Riddle Equine Hospital,
Lexington,
USA

Corinne Raphel Sweeney, DVM
Diplomate ACVIM,
School of Veterinary Medicine,
University of Pennsylvania,
USA

J.H. van der Kolk, DVM, PhD,
Diplomate EIM RNVA (Royal Netherlands Veterinary Association),
Department of Large Animal Medicine & Nutrition,
Faculty of Veterinary Medicine,
Utrecht University,
The Netherlands

Roel A. van Nieuwstadt, DVM, PhD, Diplomate RNVA (Royal Netherlands Veterinary Association),
Department of Large Animal Medicine and Nutrition,
Utrecht University,
The Netherlands

Elaine D. Watson, BVMS, MVM, PhD, FRCVS,
Department of Veterinary Clinical Studies,
University of Edinburgh,
Scotland

Preface

Presented here are over 230 questions and answers that cover current information on a wide and interesting range of the more common, and some of the less common, equine medical disorders.

Cases have been contributed by specialists in equine medicine from the United States and Europe. As we read and compared the material, it became clear that 90% of equine medical disorders are common to most countries and that, given the frequency and distances that horses travel, it is important for veterinarians to be familiar with the other 10%.

Our review presents the cases in the form of problems to be solved, given in random order, just as they may present in practice. The problems are designed to stimulate readers to make their own differential diagnoses and appropriate treatment plans. Immediately following each question we provide an answer and, perhaps more importantly, an explanation of the case. To help the reader we have also provided a list of cases classified by broad subject, an abbreviations list and a detailed index.

We take this opportunity to thank the contributors for the cases they have provided, and Manson Publishing for the speedy publication of the book. Finally, we thank you for reading the book and hope that the information serves as a useful update and review of equine medicine.

Thomas J Divers, DVM, Diplomate ACVIM
Diplomate ACVIM
Cornell University
College of Veterinary Medicine
Ithaca, NY, USA

Tim Mair, BVSc, PhD, MRCVS
Bell Equine Veterinary Clinic
Maidstone, Kent, UK

Acknowledgements

The authors are grateful to Williams & Wilkins for permission to publish **101b**, **174** and **182a** (from *Equine Diagnostic Ultrasonography* by Rantanen and McKinnon, in preparation), to Mosby–Year Book for permission to publish **24** (from *Atlas of Equine Ultrasonography* in preparation), to W.B. Saunders for permission to publish **239** (from *The Horse: Diseases and Clinical Management* by Kobluk, Ames and Geor), and to Dr Corrie Brown, DVM, PhD, DipACVP, for Figure 85.

Broad Classification of Cases

Listed are the questions and answers that deal with particular topics.

Eyes, 19, 41, 56, 74, 111, 116, 119, 122, 147, 151, 167, 176, 209, 213, 221, 230

Alimentary tract, 1, 2, 13, 18, 20, 21, 29, 40, 44, 50, 51, 52, 59, 76, 77, 78, 83, 86, 87, 92, 95, 96, 105, 108, 109, 116, 118, 124, 125, 128, 133, 144, 145, 148, 152, 162, 165, 180, 189, 207, 214, 217, 223, 227, 234

Respiratory tract, 5, 8, 10, 25, 26, 33, 38, 39, 42, 44, 54, 62, 68, 72, 113, 115, 126, 130, 131, 132, 135, 140, 145, 146, 151, 153, 170, 179, 195, 200, 212, 222, 224, 228, 235

Cardiovascular system, 22, 26, 53, 75, 102, 107, 116, 132, 158, 174, 184, 203, 205, 208, 226, 239

Liver, 34, 84, 98, 117, 169, 184, 193, 211, 225, 232

Reproductive system, 4, 7, 12, 23, 37, 42, 48, 57, 67, 70, 80, 100, 116, 121, 143, 156, 166, 181, 188, 198

Urinary tract, 11, 28, 49, 90, 97, 99, 101, 116, 120, 142, 153, 161, 177, 183, 194, 199, 231

Skin, 36, 41, 69, 71, 114, 130, 139, 163, 164, 190, 210, 236

Nervous system, 1, 5, 6, 9, 17, 55, 58, 64, 89, 110, 116, 123, 133, 137, 149, 154, 155, 157, 176, 185, 215, 237

Endocrine system, 3, 22, 26, 39, 47, 65, 88, 112, 134, 136, 143, 164, 194, 211, 233

Haematopoietic and immune systems, 3, 14, 15, 32, 34, 41, 44, 45, 49, 66, 79, 81, 106, 130, 151, 186, 193, 201, 219, 220, 225, 238

Infectious diseases, 1, 2, 3, 7, 19, 21, 23, 24, 30, 31, 35, 46, 50, 54, 55, 59, 61, 63, 66, 70, 73, 80, 83, 84, 85, 88, 91, 92, 93, 103, 109, 125, 127, 130, 131, 133, 138, 139, 140, 146, 149, 152, 159, 165, 168, 169, 171, 172, 175, 176, 180, 187, 188, 189, 190, 195, 197, 204, 209, 212, 218, 224, 229

Parasites, 16, 27, 36, 50, 69, 71, 124, 129, 150, 173, 175, 192, 206, 223

Foals, 1, 2, 7, 13, 20, 21, 23, 24, 29, 43, 44, 50, 54, 60, 61, 79, 84, 86, 95, 97, 101, 108, 116, 130, 131, 132, 137, 154, 155, 157, 159, 160, 161, 176, 178, 182, 191, 196, 197, 200, 201, 213, 215, 220, 223, 230, 237

1 & 2: Questions

1 A 23-day-old foal presents with dysphagia, a stilted gait and muscle tremors. The tail and tongue tone are weak. The vital signs and blood analyses are normal.
i. What diagnostic procedure can be used to best explain the reason for dysphagia?
ii. Assuming the condition might be caused by an infectious disease, is vaccination available as a preventive measure?
iii. How might this clinical condition be acquired?
iv. What treatments are available?

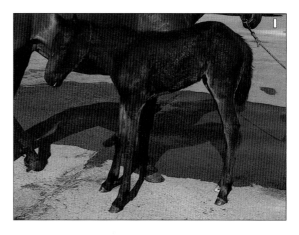

2 You are called to examine a one-day-old foal that has diarrhoea. The foal appeared normal at birth but became lethargic, depressed and developed diarrhoea (**2**). Within several hours, the diarrhoea had become haemorrhagic and the foal's clinical condition (heart rate, respiratory rate, colour and moisture of mucous membranes, lethargy and frequency of diarrhoea) had deteriorated.
i. What cause(s) of diarrhoea do you suspect in this foal?
ii. What diagnostic steps should be taken to prove your presumptive diagnosis?
iii. What preventive measures could you propose?

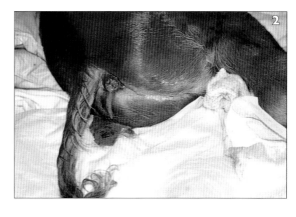

1 & 2: Answers

1 i. Endoscopy. The procedure should reveal no mechanical obstructions of the pharynx or oesophagus. Pharyngeal paresis is present in this foal. This finding, along with the presence of decreased tongue tone, is consistent with a neuromuscular disorder.
ii. Yes. These clinical signs are consistent with botulism. A vaccine has been available in the USA for Type B for a number of years, and a multivalent vaccine is being developed for commercial use.
iii. Botulism in foals is most likely the result of the toxicoinfectious form, whereby the actual organism is present in the gastrointestinal tract and the toxin is being absorbed through intestinal sites. The disease can also be acquired by infection of anaerobic wounds and by ingestion of preformed toxins (in feedstuffs such as silage).
iv. Botulism antisera can be given to protect unaffected myoneural junctions. The antibiotics of choice are aqueous forms of penicillin. Nursing care and supportive care are essential for survival.

2 i. Clostridial agents have been documented as causing severe, generally fatal diarrhoea of neonatal foals. Clostridial diarrhoea is often haemorrhagic. *Clostridium perfringens* types A, B and C have been associated with enteritis, colic and death in foals. The disease appears to be sporadic, rapidly progressive and generally fatal. Affected foals are usually less than seven days old, and signs are often seen in one-day-old foals. In the USA, *Clostridium difficile* was initially associated with haemorrhagic diarrhoea and necrotizing enterocolitis in four foals, all less than three days old. Subsequently, *Cl. difficile* and its toxin(s) have been identified in faeces of foals of various ages with mild to moderate diarrhoea. Thus, the spectrum of disease for *Cl. difficile* may be broader than that of *Cl. perfringens*.
ii. Diagnosis of *Cl. perfringens* is based on isolation of the organism and demonstration of toxin(s) in faeces or intestinal contents. This can be difficult, and only certain laboratories provide this diagnostic service. Unlike *Cl. perfringens*, demonstration of *Cl. difficile* in faeces appears sufficient to attribute diarrhoea to the organism because shedding by asymptomatic foals appears to be rare.
iii. Conclusive evidence of effective methods for prevention are lacking. One farm may have had cases in sequential years. Veterinarians at that farm administered types C and D toxoid for ruminants to pregnant mares and no ensuing cases occurred. The benefit of administering toxoid for ruminants is unclear and speculative. No evidence exists that administering antitoxin developed for sheep is beneficial.

3 This two-year-old Quarter-horse filly (**3a**) presents with an acute onset of severe swelling of the head and distal limbs. She is slightly depressed, with normal vital signs. The swelling developed suddenly, two weeks after spontaneous drainage of bilateral submandibular abscesses had occurred. Many horses on the farm were exhibiting signs of purulent nasal discharge, fever, depression and submandibular abscessation. This filly had received no treatment for the lymph node abscessation.

i. Upon closer inspection (**3b**) you notice petechial to ecchymotic haemorrhages, as well as areas of cyanosis, on the muzzle, and multiple petechial haemorrhages on the oral mucous membranes. What pathophysiologic process is most likely responsible for these signs, and what is the most likely aetiology of this process in this particular individual?
ii. Assuming that your primary differential diagnosis is correct, what would be the two most important components of your treatment plan for this filly?

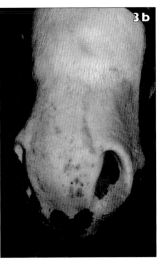

iii. What are two reasons, in addition to elimination of the *Streptococcus equi* pathogen, why antimicrobial treatment is warranted in this filly?
iv. Would you expect this filly's haemogram to have any abnormalities?

4 A six-year-old Thoroughbred mare is presented for artificial insemination with frozen semen from a Warmblood stallion. She is in oestrus and has two large pre-ovulatory follicles on her right ovary – one is 35mm and the other 38mm in diameter. At 40 days, unicornuate twins are diagnosed – how would you manage her pregnancy?

3 & 4: Answers

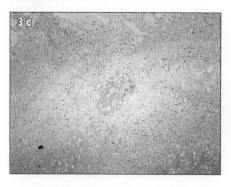

3 i. Immune-mediated vasculitis due to purpura haemorrhagica, secondary to *Str. equi* infection (strangles). The precise aetiology of this condition is uncertain, but it probably represents a hypersensititvity reaction to streptococcal antigens. Equine purpura haemorrhagica may also cause severe myositis in affected horses. The myositis, as seen in this photomicrograph of a rear limb muscle from a horse with purpura (**3c**), can be haemorrhagic and necrotizing. Marked elevations in the serum activity of muscle enzymes and myoglobinuria may be found in similarly affected horses with purpura haemorrhagica.

ii. Immunosuppressive corticosteroid treatment (eg., dexamethasone, 0.05–0.2mg/kg intravenously or intramuscularly, every 24 hours, given in the mornings) to reduce the immune-mediated vasculitis (the dose can be reduced once the oedema starts to resolve), and antimicrobial treatment (e.g., procaine penicillin, 20,000IU/kg) to eliminate the underlying streptococcal pathogens.

iii. To treat cellulitis that may occur secondary to skin sloughing. To treat secondary bacterial infections that may occur secondary to high dose steroid treatment.

iv. Mild to moderate anaemia is common; however, thrombocytopenia is rare.

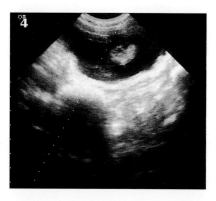

4 Management of twin pregnancies after day 35 is difficult because endometrial cups have formed, which means that it is unlikeley the mare will come into oestrus again that season if she is aborted. Furthermore, the spontaneous reduction rate after day 40 is low. Because of the high abortion rate of twins and possible long-term deleterious effects on the fertility of the mare following abortion or the birth of twins, it may be prudent to abort the mare. After day 35, daily injections of PGF-2α for 3–5 days are often necessary to induce the loss of the fetuses. One possible alternative to aborting the mare before 50 days is to attempt reduction by transvaginal ultrasound-guided needle aspiration, as shown in **4**, illustrating the embryo and needle guide (dotted line); in this case 40ml of fluid was removed and the conceptus died. However, this technique can frequently result in the loss of both conceptuses and more data are required before it can be adopted routinely. If both fetuses survive to mid-pregnancy, another method of fetal reduction is to guide a needle ultrasonically into the heart of one fetus and inject potassium chloride solution. However, this procedure needs practice and can result in the death of both fetuses.

5 & 6: Questions

5 You are asked to examine a five-year-old Thoroughbred gelding that is being used as a dressage horse. The owner tells you that the horse has gradually developed a respiratory noise during exercise over a period of 18 months. The noise is most obvious at the gallop. Exercise tolerance has not been affected.
i. What is the most likely diagnosis?
ii. How can you confirm the diagnosis?
iii. What is the aetiology of this condition?
iv. Describe the typical respiratory noise caused by this condition.
v. Is this condition likely to affect the exercise tolerance or ability of the horse?

6 A nine-year-old Quarterhorse (6) is examined because of an acute onset of trembling, standing with the hind feet abnormally positioned under the body and lying down more than normal. The horse walks normally, but when made to stand becomes restless and

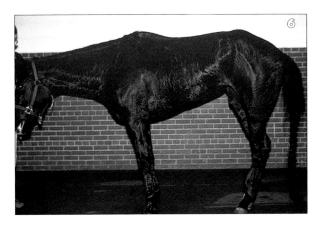

constantly shifts the weight on the hindlimbs. The horse's head appears to be held lower than normal and the tail head appears abnormally elevated. When the horse lies down, fine fasciculations are noted in many muscles. The horse has reduced muscle mass, especially in the muscles of the limbs and neck. Serum chemistry laboratory values are normal except for a creatine kinase concentration of 1260u/l and aspartate aminotransferase concentration of 740u/l.
i. What body system appears to be abnormal in this case?
ii. What samples could be taken to help confirm the diagnosis?

5 & 6: Answers

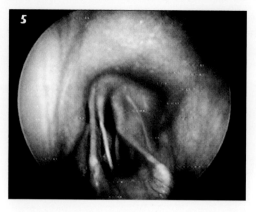

5 i. Laryngeal hemiplegia (most commonly left laryngeal hemiplegia), also known as recurrent laryngeal neuropathy.
ii. Confirmation of the diagnosis requires endoscopic examination of the larynx. In cases where the paralysis is complete the affected arytenoid cartilage exhibits little or no movement and is displaced towards the midline. In horses that have incomplete laryngeal paralysis there is a loss of abductor function which is variable depending on the degree of the underlying neurogenic muscle atrophy. In cases where there is doubt about the asymmetry of the larynx and laryngeal movements at rest, endoscopy can be performed during exercise on a high speed treadmill.

In addition to endoscopic examination, evaluation of the horse with suspected laryngeal hemiplegia may include external palpation of the larynx, the arytenoid depression test, the slap test and estimation of intermandibular width.
iii. The precise aetiology of idiopathic left laryngeal hemiplegia is uncertain. The disease occurs primarily in larger breeds of horses. The lesion in the recurrent laryngeal nerve is characterized by a distal, progressive loss of large myelinated nerve fibres. The nerve damage results in neurogenic atrophy of the associated intrinsic muscles of the larynx, especially the cricoarytenoideus dorsalis muscle, which is responsible for abduction of the arytenoid cartilage.

In a small number of cases there is an identifiable cause of acquired left (or right) laryngeal hemiplegia. Trauma, inflammatory processes (eg. strangles), tumours and surgical procedures affecting the left (or right) recurrent laryngeal nerve, and some intoxications can all cause nerve damage. Bilateral laryngeal paralysis is rare, and is usually the result of CNS disease.
iv. The typical respiratory sound is a high-pitched, inspiratory noise described as a 'whistle' or 'roar' that occurs during canter and gallop. Sometimes, the ability to vocalize is impaired.
v. The effect of left laryngeal hemiplegia on exercise tolerance is debatable. The effect will depend on the degree of paresis or paralysis, and the type of exercise involved. Significant loss of cricoarytenoid muscle function may lead to partial or even total laryngeal collapse during strenuous exercise (5).

6 i. The weakness, muscle wasting, fasciculations, tremors and elevated muscle enzymes suggest that the muscle and/or the neuromuscular system is affected. The normal appetite and normal gait make intestinal and skeletal disorders unlikely.
ii. A muscle biopsy would help support the diagnosis. This could be taken from the dorsal tail muscle (sacrococcygealis dorsalis). See question **17**.

7 & 8: Questions

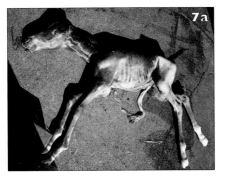 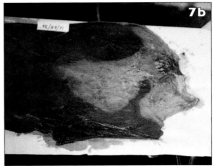

7 A mare aborts on a stud farm at 10 months of gestation. The fetus (**7a**) and the placenta (**7b**) are shown. What might be the cause of the abortion?

8 The endoscopic appearance of the trachea of a three-year-old Thoroughbred colt examined approximately one hour following racing is illustrated in **8a**.
i. What is the likely diagnosis?
ii. This condition is present in what regions of the world?
iii. What is the approximate frequency of this condition?
iv. A cell type commonly found in the tracheal secretions of horses with this condition is illustrated in **8b**. What is the cell in the middle of the picture?

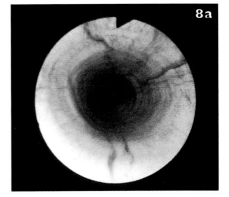

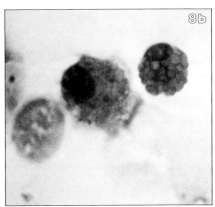

7 & 8: Answers

7 Causes of abortion can be divided into three categories: infectious, non-infectious and unknown. Between 16 and 40% of abortions have been reported to be of unknown causes. Approximately 16% are thought to be infectious and 48% non-infectious. In this case the aborted fetus looks emaciated. Crown–rump length of the fetus should always be measured to give an indication of normality of growth. Small fetuses are often associated with chronic placental insufficiency resulting from an inflammatory or non-inflammatory placental lesion.

The placenta shows evidence of a thickened distinct area of placentitis at the cervical area which is clearly demarcated from the rest of the placenta. The distribution of the lesion indicates an ascending infection from the vagina, which appears to be the most common route. Bacteria and fungi isolated from fetal membranes may or may not be the primary cause of abortion. It is not unusual for aborted fetuses to be infected with a mixed population of bacteria and fungi. Many organisms found in aborted material are opportunistic pathogens of adult horses and are common in the stable environment. Therefore, microbiological findings should be interpreted along with the findings of the pathological examination. The most common bacteria isolated are *Streptococcus zooepidemicus*, *Escherichia coli* and *Pseudomonas aeruginosa*. *Leptospira* spp. are now being isolated and *Ehrlichia* spp. may be implicated. The most common fungi isolated are *Aspergillus fumigatus* and *Absidia* spp.

8 i. Exercise-induced pulmonary haemorrhage (EIPH) occurs when horses have blood or other evidence of haemorrhage into an airway after strenuous exercise.
ii. EIPH occurs worldwide wherever there are horses engaged in strenuous exercise.
iii. EIPH occurs at about the same frequency worldwide. The prevalence rates vary depending on the means of diagnosis. Studies determining EIPH based on blood at the nares have the lowest percentage (less than 3%). Studies determining EIPH based on endoscopic findings find prevalence rates up to 75%. Studies based on the presence of haemosiderin-laden macrophages in bronchoalveolar lavage samples show a prevalence rate of up to 100%. It is commonly thought that virtually all horses engaged in strenuous exercise bleed in the lungs.
iv. Macrophages laden with haemosiderin, often called haemosiderophages. While these cells are very sensitive indicators of pulmonary haemorrhage, they are not specific as to the cause of the haemorrhage; for example, haemosiderophages can be found in tracheal aspirate from a horse with a haemorrhagic pneumonia.

9–11: Questions

9 A seven-day-old Warmblood filly is found recumbent and unable to rise. The animal had showed a stiff gait during the previous day. The heart rate is increased (92bpm) and body temperature is within normal limits.
i. What is the most likely metabolic disease?
ii. How could you confirm the diagnosis?
iii. Do you think it is valuable to take a blood sample for assessment of the glutathione peroxidase activity 10 days after selenium administration to monitor its effect?
iv. Can you describe the oral vitamin E absorption test?
v. How would you prevent the occurrence of more cases?
vi. What potential complication may develop from selenium supplementation?

10 A five-year-old Thoroughbred gelding is presented with a history of progressively worsening exercise intolerance. A harsh, inspiratory respiratory noise is heard at the canter and gallop. Endoscopy reveals markedly reduced arytenoid movements and deformity (**10**).
i. What is your diagnosis?
ii. What is the aetiology and pathogenesis of this condition?
iii. How can this condition be treated?

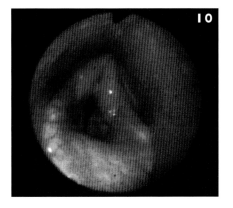

11 A seven-year-old Standardbred is examined because of weight loss and partial anorexia. There are no significant findings on clinical examination except for severe dental tartar and oral erosions (**11**).
i. Failure of which organ system might cause all of these clinical signs?
ii. What other laboratory test would confirm the diagnosis?

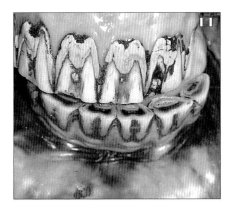

9–11: Answers

9 i. Nutritional muscular dystrophy, often associated with steatitis. This disorder of neonatal foals has a very poor prognosis. A more chronic form can be seen in foals around the time of weaning. The ligamentum nuchae is usually swollen as well as adipose tissue in the inguinal regions. There is a reluctance to walk, sluggishness and increased respiratory and pulse rates.

ii. Increased activities of creatine kinase, aspartate aminotransferase and lactate dehydrogenase are the most commonly used laboratory aids in the diagnosis of nutritional muscular dystrophy. In addition, the glutathione peroxidase activity in the foal reflects the quantity of selenium given to the dam. In general, the glutathione peroxidase activity is a reliable indicator of dietary selenium intake and the response to selenium supplementation. Selenium protects against peroxidase damage to cellular membranes, thereby acting together with vitamin E.

iii. No. The enzyme glutathione peroxidase contains four atoms of selenium, and is located in the erythrocyte. Selenium is incorporated into erythrocyte glutathione peroxidase only during erythropoiesis. Equine erythrocytes have an average life span of about 150 days, and an increase in enzyme activity of the blood will not occur within four weeks after administration.

iv. The oral vitamin E absorption test is performed by administering 100 IU of alpha-tocopherol (vitamin E) added to 1kg of concentrate feed after food has been withheld for 12 hours. Blood samples are collected 12 hours after the ingestion of the test dose and just prior to the administration of it. In normal horses the plasma alpha-tocopherol concentration doubles at 12 hours.

v. Both selenium and vitamin E should be provided to other horses on the farm, and especially to pregnant mares as selenium crosses the placenta. Daily oral administration of 0.06g/kg bodyweight of a supplement containing about 105,000 IU vitamin E/kg and 15mg selenium/kg is necessary.

vi. Selenium toxicosis reflected by two forms, namely 'blind staggers' and 'alkali disease'.

10 i. Arytenoid chondritis (arytenoid chondropathy). There is asymmetry of the larynx and an intraluminal mass of granulation tissue is seen projecting from the luminal surface of the corniculate process.

ii. The aetiology and pathogenesis are uncertain. The disease is seen most commonly in young Thoroughbreds. One or both arytenoids may be affected. Infection and trauma have been suggested as possible aetiologies. The cartilage is chronically inflamed, and the normal hyaline cartilage is replaced by granulation tissue infiltrated by neutrophils, lymphocytes and macrophages.

iii. A prolonged course of antibiotics with systemic anti-inflammatory therapy is occasionally successful, but recurrence of clinical disease may occur on cessation of therapy. Surgical therapy involves total, partial or subtotal arytenoidectomy.

11 i. Renal failure caused all of the clinical signs in this case.

ii. An elevation in serum creatinine and urea with isosthenuric urine (1.008–1.014) would confirm the diagnosis.

12 What are the common causes of failure to show oestrus behaviour in non-pregnant mares?

13 You are asked to examine a 24-hour-old colt foal that has started to show signs of colic within the past hour. The owner reports that the foal was born at term, with no complications, and has been observed to suckle vigorously during the first day of life (**13**).

i. How must your routine examination of a horse with colic be adapted to treat the foal?
ii. What additional diagnostic aids may be useful in this situation that are not applicable to examination of intestinal problems in larger horses?

14 A seven-year-old pastured pleasure horse is examined for depression and red urine. Abnormalities on clinical examination include: severe tachycardia (92bpm), tachypnoea (44bpm) and discoloured mucous membranes (**14**). Abnormal laboratory findings include: pink plasma; PCV, 9%; MCV, 68fl; neutrophilia and mild azotaemia. Heinz bodies were seen on some RBCs. List differentials for this condition.

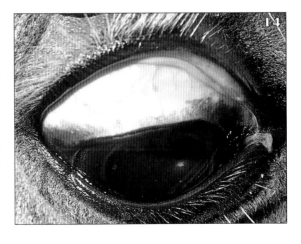

15 How would you recognize the following in a blood smear:
i. Howell–Jolly bodies?
ii. Heinz bodies?

12–15: Answers

12 In mares which fail to show oestrus behaviour it is important to distinguish between mares with a psychic disorder which have normal ovarian cyclicity, and true anoestrus mares which are not cycling. Another category of mares which will fail to show oestrus contains those that have prolonged luteal activity. Prolongation of dioestrus is a major cause of infertility in the mare and can be due to:
- Repeated ovulation late in dioestrus when the corpus luteum (CL) resulting from this ovulation will fail to respond to uterine PGF-2α.
- Inability of the uterus to produce PGF-2α, such as in mares with pyometra.
- Idiopathic persistence of the CL from failure to release PGF-2α.

13 i. Evaluation of the colicking foal is fraught with difficulties, both technical and interpretational. The first difficulty is interpretation of the degree of pain exhibited. This takes the form of an all-or-nothing response giving the clinician little opportunity to evaluate the severity of pain. Cardiovascular parameters, so important in revealing endotoxaemia in the older animal, are not a reliable indicator in the neonate. This may be due to immaturity of the physiological response to endotoxins. The pivotal technique of palpation per rectum is limited to digital examination of the terminal few centimetres of rectum thus denying the clinician the opportunity to evaluate directly a large proportion of the abdomen and its contents. Abdominocentesis can be difficult in the foal which can usually be restrained in lateral recumbency only. However, other aspects of the routine gastrointestinal examination are possible, eg. auscultation of the abdomen and passage of a nasogastric tube.
ii. Both radiography and ultrasonography can be utilized to image the contents of the abdomen and to explore for abnormalities. Intussusceptions can be visualized directly by ultrasonography. Concreted colonic contents may be observed on radiographs as can gas and fluid distension of intestine proximal to an obstruction. Radiography (readily performed with the foal in lateral recumbency) can also be used to detect intestinal distension. The diagnosis of retained meconium and its differentiation from atresia coli may be aided by radiography, in particular the use of barium enemas to delineate the rectum and terminal colon.

14 The differentials should include red maple toxicosis, onion poisoning and other regional toxins that might cause oxidative damage to the red blood cells.

15 i. Howell–Jolly bodies are round basophilic nuclear remnants seen near the periphery of erythrocytes.
ii. Heinz bodies represent precipitation of denatured haemoglobin seen in Wright–Giemsa stained blood smears as protrusions of the erythrocyte membrane from the surface of the cell or as pale blue bodies within the cell.

16 & 17: Questions

16 Two cases of fatal diarrhoea have occurred in mature adult horses at a riding stables. On the basis of identification of coiled cyathostome larvae within large intestinal mucosa (**16**) these cases have been diagnosed as larval cyathostomosis. Both cases were reported to have been dosed with anthelmintics at two monthly intervals throughout the two-year duration of ownership. What are the possible explanations for the occurrence of these clinical cases?

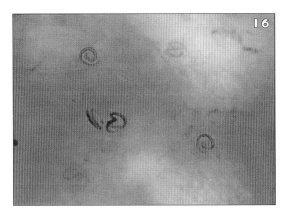

17 A seven-year-old Quarterhorse is noted to be lying down more than normal. The appetite is good and there are no cranial nerve deficits. The horse walks normally, but when made to stand becomes restless and almost constantly shifts the weight on the hindlimbs. The horse's head appears to be held lower than normal and the tail head appears abnormally elevated. When the horse lies down, fine fasciculations are noted in many muscles. The horse appears thin and has reduced muscle mass, especially in the muscles of the limbs and neck. Routine CBC and serum chemistry laboratory values are normal except for mildly elevated creatine kinase and aspartate aminotransferase concentrations.

You perform an EMG on several muscles (**17a**) and a muscle biopsy of the dorsal tail muscle (**17b**). The horse is boarded at a stable without pasture and is fed large amounts of grain and poor quality grass hay. What is the diagnosis and which vitamin deficiency has been associated with this disease?

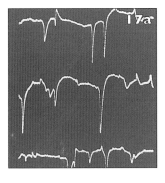

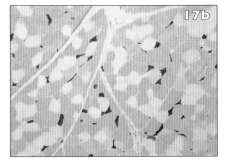

16 & 17: Answers

16 There are three possible explanations:
- The larval burdens which gave rise to the clinical cyathostomosis were acquired prior to present ownership and they have survived multiple anthelmintic doses given since purchase. It is well known that larvae present within large intestinal mucosa in a state of arrested larval development (ALD) are relatively refractory to the action of all anthelmintics and it has been shown that the state of ALD can be maintained for several years. Clinical episodes of larval cyathostomosis are most commonly identified in animals between one and five years of age during winter and spring but cases do occur in older or aged animals and also during any season.
- The parasite prophylaxis programme failed to prevent the contamination of grazing with infective cyathostome larvae. In many management situations, animals which share grazing are not maintained on a single parasite control programme, most commonly if the animals have different owners. As a result, instead of a structured programme with synchronized anthelmintic dosing of the entire grazing group, there is haphazard, random dosing and it is likely that at any point in time a proportion of animals will be passing cyathostome eggs in faeces and pasture larval contamination will be on-going. Other factors such as grazing intensity or environmental conditions may contribute to certain animals ingesting pathogenic numbers of cyathostome larvae.
- Anthelmintic resistance is common in cyathostome populations throughout the world. To date, in cyathostomes, resistance has been documented to the benzimidazole (BZ) anthelmintic compounds and pyrantel. In BZ resistance there is usually cross resistance to all the different benzimidazoles, but on occasions BZ-resistant cyathostomes remain sensitive to oxibendazole.

17 The diagnosis is equine motor neuron disease. Vitamin E (alpha tocopherol) has been consistently low in plasma, fat, nervous tissue and muscle of affected horses. The EMG depicts positive sharp waves and the muscle biopsy shows angular atrophy, mostly of type 1 fibres. These findings are characteristic of muscular denervation.

18–20: Questions

18 This stomach lesion (**18a**) was found in a four-year-old horse which was killed because of an orthopaedic problem. It shows a classic rupture of the stomach wall along the greater curvature. The haemorrhage along the rupture line can help to distinguish this ante-mortem rupture from the equally common post-mortem rupture. This seromuscular rupture had

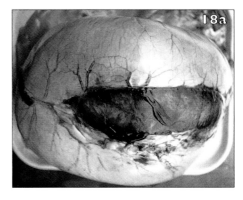

occurred without an associated rupture of the mucosa.
i. How can gastric rupture occur?
ii. What are the clinical signs associated with a complete rupture of the stomach wall?

19 The eye shown in **19** has a corneal ulcer.
i. Is this ulcer infected or not?
ii. What is your most important diagnostic test?
iii. If this ulcer was associated with a bacterial infection, how would you treat it?

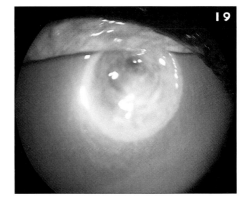

20 You are trying to determine whether a foal with acute colic needs to be managed medically or surgically. The abdominal radiograph (**20**) is obtained from the lateral position of the foal while it is standing. What is your interpretation of this radiograph?

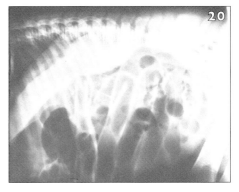

18–20: Answers

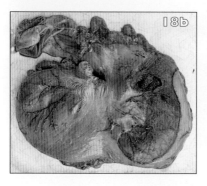

18 Gastric rupture can occur as a result of overeating, especially readily fermentable foodstuffs such as grain, grass and foodpulp. More commonly gastric rupture is associated with some form of intestinal obstruction that causes small-intestine contents to reflux into the stomach, where they accumulate and cause distension. The horse has a powerful lower oesophageal sphincter that prevents vomiting. The distension of the stomach may therefore become progressive, resulting in a rising intragastric pressure that ultimately may lead to rupture. Examples of obstructions that can result in gastric rupture include mechanical obstructions of small intestine, paralytic ileus of the intestines (eg. secondary to peritonitis or enteritis, or post-operative ileus), grass sickness (in the UK), intra-abdominal abscesses and adhesions. Shown in **18b** is a ruptured stomach of a horse with rupture of the seromuscular layers only and a smaller area with additional rupture of the mucosa.
ii. Signs of gastric rupture are usually preceded by severe pain, increased heart rate, distended abdomen and sometimes retching. Cyanosis may be present. Once the stomach ruptures, the signs of pain may disappear, but the horse develops signs related to acidosis, toxaemia and peracute peritonitis, causing severe depression, sweating and anxiety. Death follows within a few hours.

19 i. Infected. There are necrotic edges, a cratered base, diffuse corneal oedema, severe injection and severe pain.
ii. A corneal scraping from the necrotic edges of this ulcer should be obtained for Gram staining and for culture. Gram-stained scrapings are the most immediate aid to help differentiate Gram-positive, Gram-negative and fungal organisms, thereby allowing appropriate selection of therapy.
iii. Aggressive antibiotic therapy is indicated. Topical antibiotics should be applied frequently, preferably every hour initially. This may be most effectively achieved using a subpalpebral lavage catheter. After the first 12–24 hours, the rate of medication may be reduced to every 4–6 hours. Selection of antibiotics should be based on the results of cytology, and confirmed by culture and sensitivity testing. Topical atropine is indicated to relieve discomfort if miosis and ciliary spasm are present.

The ulcer should be carefully re-evaluated within 24 hours of initiation of treatment, and treatment modified as necessary.

20 Although radiographic findings rarely provide a definitive answer as to whether or not surgery is indicated for foals with colic, the presence of multiple erectile loops of distended small intestine is most consistent with a mechanical obstruction of the small intestine. These findings would support the decision to manage this foal surgically. An impaction of the ileum was detected in this foal.

21 The foals presented in **21a** and **21b** are exhibiting clinical signs of colic, bruxism, salivation and intermittent nursing. Laboratory values vary and are not diagnostically consistent for the disease process.
i. What is the clinical condition?
ii. How can the diagnosis be confirmed?
iii. What four forms of the syndrome exist?
iv. What are the types of treatment available?

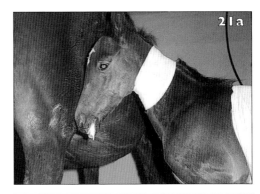

22 An eight-year-old Shetland pony mare is presented with a six-hour history of suspected colic (**22**). The mare had foaled four months previously. The pulse rate is 72/minute and synchronous diaphragmatic flutter is noticed. The plasma calcium concentration is decreased.
i. What is the likely diagnosis?
ii. During which phase of the ECG does synchronous diaphragmatic flutter occur?
iii. What are the other causes of hypocalcaemia in the horse?
iv. How many parathyroid glands does a horse possess?

21 & 22: Answers

21 i. The gastric ulcer syndrome.
ii. By endoscopic examination of the stomach.
iii. The four forms are:
- Silent ulcers. These are foals which have never exhibited clinical signs of gastric ulceration, but have active ulceration shown by endoscopy or as an incidental finding at necropsy examination in foals that have died from other causes.
- Clinical ulcers. These foals may exhibit any of the clinical signs of ulceration. Colic, frequently with prolonged periods of lying on their back, anxiety when nursing (restlessness or tail-flagging, milk over the muzzle area), bruxism (intermittent or persistent), refluxing, excessive salivation and 'drooling', a history or intermittent presence of diarrhoea. Oral ulcers have been noted but are considered inconsistent as a clinical sign.
- Delayed gastric emptying secondary to gastric dysfunction, a pyloric stricture, duodenitis or a combination of these lesions.
- Peritonitis secondary to ulcer perforation.

iv. Treatments consist of stomach buffers, coating medications, acid receptor antagonists, acid inhibitors, upper gastrointestinal motility stimulants and surgery. Antibiotics are used subjectively since *Heliobacter pylori* has not been confirmed in equids. The preferred medications are the use of sucralfate to bind to and protect the ulcer site, alone or in combination with an H_2 receptor antagonist to prevent acid formation (eg. cimetidine, ranitidine). Omeprazole can be used as an effective H^+ proton inhibitor. The use of prostaglandin E is regarded as preventive, and is not currently clinically advocated. Bypass surgery is recommended for foals with persistent gastric reflux that have not responded to the motility stimulants (metoclopramide or cisapride). Successful surgical bypass individuals may be subject to growth retardation.

22 i. Lactation tetany. Lactation tetany involves mares usually in the first 10 days of lactation and sometimes a few days post weaning. The initial clinical signs are a stiff, stilted gait, muscle fasciculations, especially of the triceps muscles, tachycardia and synchronous diaphragmatic flutter. Parathyroid hormone concentrations can be within the normal range.
ii. Synchronous diaphragmatic flutter is a clinical sign in which contraction of the diaphragm is synchronous with atrial depolarization.
iii. Hypocalcaemia can also be associated with diarrhoea, excessive sweat loss in endurance horses, stress due to transport, blister beetle toxicosis, oxalate toxicity and sometimes primary hypoparathyroidism.
iv. The parathyroid glands consist of two pairs in the horse. One pair is located on the mediodorsal surface of the thyroid glands. The other pair is located near the bifurcation of the bicarotid trunk at the level of the first rib.

23–25: Questions

23 The abortion on a stud farm at 10 months of gestation shown in **7a** and **7b** needs to be assessed.
i. What samples are you going to take to diagnose the cause?
ii. What tests need to be performed to attempt to identify the cause of the abortion?
iii. Do any precautions need to be taken on the stud farm in the short term?

24 A sonogram of the urachus (**24**) was obtained from a three-week-old foal with a soft swollen umbilical stump.
i. What is the diagnosis?
ii. What are the most frequent organisms involved?
iii. What options are there for treatment of this problem?

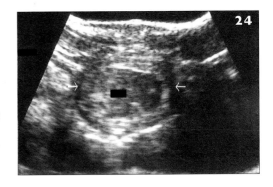

25 An eight-year-old pony mare collapsed two minutes after receiving an intramuscular injection of procaine penicillin. The mucous membranes are pale and cyanosed, and the pulse is rapid (heart rate 70bpm) and weak. The pony is tachypnoeic and dyspnoeic with widespread crackling lung sounds, and frothy fluid drains from the nose (**25**). After about one minute, the pony dies.

i. What is the most likely diagnosis?
ii. How could you have attempted treatment if the pony had survived long enough?
iii. What other adverse reaction is more commonly associated with administration of procaine penicillin?

23–25: Answers

23 i., ii. The fetus plus the placenta should be sent as soon as possible to a recognized laboratory for post-mortem examination. At the laboratory, samples will be taken for microbiological culture, virology and histological examination. Any gross pathology of the placenta or fetal organs should be noted.
- Microbiology: samples collected aseptically from the placenta, liver, stomach contents, lung and kidney.
- Virology: samples stored chilled or frozen from the placenta, spleen, liver, lymph nodes, adrenal, lung and thymus for virus isolation ± fluorescent antibody staining.
- Histology: samples fixed in 10% buffered formalin or Bouin's solution from the placenta, spleen, liver, lymph nodes, adrenal, kidney, lung, thymus and heart.

Two serum samples should be collected from the mare at an interval of at least two weeks and sent to a recognized laboratory for antibody titres against EHV and EVA.
iii. The mare should be placed in strict isolation, pending laboratory results. All bedding should be disinfected, left for 48 hours and then burned. The stable should be thoroughly disinfected and steam cleaned. No horses should be permitted to leave the stud, nor any pregnant mares brought onto the stud, until the possibility of EHV or EVA infection is excluded. If herpesvirus is isolated all owners of other horses at the stud should be notified.

24 i. Urachal abscess.
ii. Beta haemolytic *Streptococcus* sp. is probably the most common, often accompanied by Gram-negative organisms, particularly *Eschericia coli*, *Staphylococcus aureus* and anaerobes may also be cultured.
iii. Surgical excision is not always necessary. Medical treatment is certainly appropriate if only the urachus is involved and it is not an excessively large abscess. If there is drainage from the stump, drainage should be allowed to occur, rather than applying topical irritants. The age of the foal, blood work and presence of fevers are also important variables used in deciding appropriate management. In this case the soft umbilical swelling (which was determined to be a pocket of pus contiguous with the urachus on sonographic evaluation) was lanced. The foal was placed on an oral antimicrobial and the stump kept clean. The foal recovered without incident. A follow-up sonogram, performed several days after the drainage ceased and the stump appeared normal, revealed umbilical remnants of normal appearance and no evidence of residual infection.

25 i. Pulmonary oedema and shock caused by anaphylactic or anaphylactoid reaction to penicillin.
ii. Epinephrine (3–5ml of 1:1000 solution diluted in 20–30ml of saline and given slowly i/v), dexamethasone (0.25mg/kg i/v), frusemide (furosemide) (1.0mg/kg) and intranasal oxygen.
iii. Procaine toxicity characterized by hyperexcitability, circling, snorting and, occasionally, collapse and seizures.

26–28: Questions

26 A four-year-old filly is presented with a history of depression and ventral swelling which has developed over the previous month. On physical examination the main findings are a rapid, regular but weak pulse, marked jugular distension and ventral oedema, particularly in the pectoral region (26). On auscultation the heart was not audible.
i. What are the differential diagnoses?
ii. How can these possibilities be investigated?

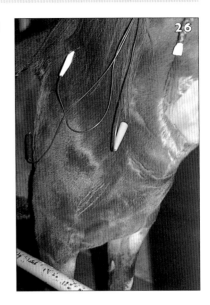

27 What is the easiest way of assessing the presence of anthelmintic resistance of cyathostome species in grazing horses?

28 Shown (28) is the sonogram of the kidney of a horse with chronic renal failure.
i. What are the abnormalities noted?
ii. These abnormalities were found bilaterally. What is the most likely reason for these abnormalities?

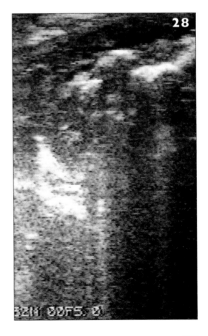

26–28: Answers

26 i. Major differential diagnoses in a horse with pectoral oedema and jugular distension are pericarditis and mediastinal lymphosarcoma or other mediastinal neoplasia. In pericarditis the heart sounds are muffled, whereas with a mediastinal mass the cardiac silhouette may be displaced caudally. Complex congenital cardiac conditions such as tricuspid atresia could present in this way; however, these lesions would not be considered in an adult horse. Right-sided heart failure, for example in right ventricular cardiomyopathy, or congestive heart failure can cause these signs. However, the degree of oedema present is more severe than is normally seen in adult horses with congestive heart failure; also, in both congenital cardiac conditions and with valvular disease the heart would be audible and loud murmurs would be expected. Pleuropneumonia can cause pectoral oedema but would not normally be associated with jugular distension. Equally, an abscess in the pectoral region, for example an injection abscess, causes pectoral oedema but not jugular distension.
ii. Thoracic ultrasonography can be used to demonstrate pericarditis and rule out other cardiac conditions. It is also useful to demonstrate mediastinal masses. Cytological examination of pleural fluid is indicated if thoracic neoplasia is suspected.

27 In situations of cyathostome-associated disease in animals known to have undergone parasite prophylaxis using anthelmintic dosing, it is essential to investigate for the presence of anthelmintic resistance. The most practical method is to perform a faecal egg count reduction test (FECRT), the basis of which is to compare faecal worm egg counts (FWECs) prior to, then 10–14 days following, anthelmintic dosing. For modern anthelmintics there should be at least an 85% reduction of FWECs following dosing. To obtain meaningful results it is necessary to include only animals with a pre-test positive FWEC. Ideally, the test should be performed by comparison of data from a negative (untreated) control group and several groups given different classes of anthelmintic compounds. Practically, it may not be possible to subdivide animals in this way; an initial observation of a high proportion of animals with positive FWECs following anthelmintic dosing suggests drug resistance, which should then be investigated in more detail. Following the development of anthelmintic resistance in a population of cyathostomes, the use of that compound (or group of compounds) should be excluded indefinitely from the parasite prophylaxis programme.

28 i. Calculi can be seen in the kidney, casting shadows.
ii. The most likely cause is renal crest necrosis (non-steroidal anti-inflammatory toxicosis) and mineralization of the sloughed papillae.

29–32: Questions

29 List, in order of likelihood, the differential diagnoses that you would wish to rule-in/rule-out when assessing a 24-hour-old colt foal that has developed signs of colic.

30 Shown (30) is the foot of a horse affected by laminitis that has had a dorsal wall resection (dorsal wall stripping). What are the indications for performing this procedure?

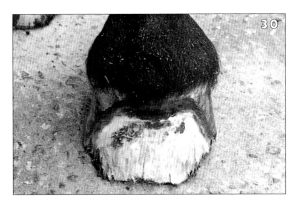

31 A three-year-old horse is examined because of an acute onset of fever and oedema of the limbs. Examination of the mucous membranes reveals the abnormality seen in the mouth in **31**. The horse was exposed to strangles and had a purulent nasal discharge three weeks earlier. CBC reveals a neutrophilia and mild anaemia (PCV, 28%). The platelet count is normal and serology for EIA, EVA and *Erhlichia equi* are negative.

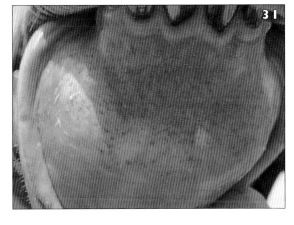

What is the most likely diagnosis in this case?

32 A transfusion is to be given to the horse (500kg) described in **14**.
i. How much blood will be needed to raise the PCV to 20% if the PCV of the donor is 45%?
ii. How much blood can be safely taken from the 500kg healthy donor?
iii. What red cell antigens and antibodies would be least desirable for an equine donor?

29–32: Answers

29 In order of likelihood the differential diagnoses are:
- *Meconium impaction.* The age and sex of the foal are strongly suggestive of this problem which is by far the most common cause of colic in animals of this age. Diagnosis is made by a suggestive history (failure to observe meconium being passed), physical examination (palpation of meconium pellets in rectum) and diagnostic enema. Radiography may be useful in rare cases where doubt still exists after performing these procedures.
- *Intestinal obstruction.* This is much less likely than retained meconium but should be considered if physical examination reveals the normal passage of meconium (brown/green) and foal faeces (pasty yellow). Foals can develop any of the different types of obstruction that adults can. Intussusceptions of small intestine are more common in foals than in older horses and the possibility of intestinal entrapment in a hernia (umbilical, inguinal, scrotal or diaphragmatic) should also be considered.
- *Congenital abnormality.* Atresia coli is the congenital non-development of various segments of the colon, and this can lead to colic within a similar time scale to, and with similar signs as, meconium impaction. Fortunately, it is extremely uncommon. Contrast radiography may be useful diagnostically but exploratory laparotomy is necessary to confirm the diagnosis. Ileocolonic aganglionosis has been described in all-white progeny of Overo Paint horses (lethal white syndrome) and myenteric hypoganglionosis has been reported. Both of these latter conditions are extremely rare but could present with the signs described above.

30 Absolute guidelines relating to when to perform dorsal wall stripping do not exist. Indications for the procedure include cases of laminitis with excessive distal phalangeal rotation (more than 8–10°) and sinking, slight rotation and sinking that increase with time, submural seromas and gas pockets, oozing of serum at the coronary band and submural sepsis. In some cases where these features are not recognized, it may be helpful to remove a small section of the dorsal wall and examine the underlying laminar tissue before deciding whether or not to proceed with more radical stripping.

31 The most likely diagnosis is purpura haemorrhagica.

32 i. (20% – 9%/45%) x (500kg x 8%) = 10 litres.
ii. 15–20% of the blood can be safely removed from a healthy horse or 6–8 litres from a 500kg horse.
iii. Aa and Qa antigens and antibodies are least desirable in a horse to be used as a donor.

33 & 34: Questions

33 You are asked to examine a three-year-old trotter mare with a history of poor performance. The trainer tells you that the animal seems to have no trouble at the beginning of the race, but after approximately three quarters of a mile the trotter suddenly falls back and appears to be choking. At the finish the trouble seems to be over; during clinical examination at rest you find no significant abnormalities.
i. What is the most likely diagnosis?
ii. Describe the typical respiratory noise caused by this condition.
iii. How can you confirm this diagnosis?

34 A 10-year-old Tennessee Walking Horse mare, kept on pasture, was presented due to a rapid onset of abnormal mentation manifested as compulsive circling, repeated yawning and apparent blindness. Her sclera were markedly injected (**34a**) and her oral mucous membranes were toxic, with an underlying icterus (**34b**). Hepato-encephalopathy was diagnosed on the basis of elevated serum liver enzyme activities, bilirubin concentration and bile acid concentration. Before performing a liver biopsy, a clotting profile was performed.

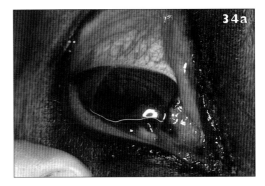

i. What does the prothrombin time (PT) test?
ii. What does the activated partial thromboplastin time (APTT) test?
iii. Which clotting time, PT or APTT, would you expect to be prolonged first in a horse with hepatic failure, and why?
iv. Horses with liver failure are often endotoxaemic due to the failure of the hepatic macrophages to phagocytoze gut-derived endotoxin. Can you explain why this endotoxaemia might exacerbate factor VII consumption?

33 & 34: Answers

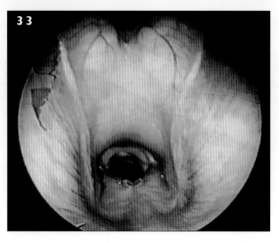

33 i. Dorsal displacement of the soft palate (DDSP).
ii. A hoarse, loud gurgling or fluttering sound that occurs during exercise and is heard during both inspiration and expiration. The noise may disappear immediately after the horse swallows, or the palate may remain displaced for the remainder of the race. Usually, by the time the horse has returned to the paddock the palate has returned to its normal position and the noise has stopped.
iii. The diagnosis of DDSP is based on the typical history of the animal, observation of behaviour and the respiratory sounds made by the animal during exercise, and endoscopic examination. Most horses with DDSP during exercise appear normal on endoscopy at rest. It is important to rule out other possible causes of upper airway obstruction. Endoscopy immediately after or during strenuous exercise (on a high speed treadmill) offers the best chance of confirming DDSP. **33** shows an endoscopic view of the pharynx and larynx of a horse with DDSP. The rostral and lateral margins of the osteum intrapharyngeum are visible, but the epiglottis is not visible. The free edge of the soft palate is ulcerated in this horse.

34 i. PT tests the extrinsic arm of the coagulation cascade, which begins with activation of factor VII.
ii. APTT tests the intrinsic arm of the coagulation cascade, which begins with activation of factor XII.
iii. The PT, because factor VII has a very short half-life. Therefore, with decreased production, the available factor VII is depleted more quickly than the factors with longer half-lives.
iv. Endotoxin stimulates extrinsic coagulation by stimulating the expression of tissue factor on the surface of mononuclear cells. Tissue factor binds factor VII, thereby initiating the extrinsic coagulation cascade.

35–37: Questions

35 Shown (35) is the sagittal section of the foot of a pony destroyed because of severe acute laminitis.
i. What is the pathophysiology of acute laminitis, and why has the distal phalanx (pedal bone) rotated in this horse?
ii. What are the 'Obel grades' of acute laminitis?
iii. What is meant by subacute, acute and refractory laminitis?

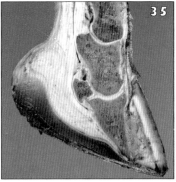

36 This horse (36) was presented in winter for a complaint of tail rubbing. Historically, the horse also rubbed its face, neck and ventrum during the summer but the pruritus in those other areas stopped at the first frost. Aside from some mild seborrhoea of the mane, the horse is otherwise normal. The pruritus stops with the administration of prednisolone.
i. List three differential diagnoses.
ii. In light of the horse's response to corticosteroid administration, which differential diagnosis can be excluded?

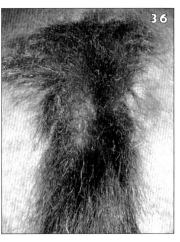

37 A four-year-old, slightly built, 15hh Thoroughbred mare is presented in August for an infertility investigation. She has been teased regularly over the summer but has failed to show signs of oestrus. On palpation per rectum she has ovaries with dimensions 1.0 x 1.0 x 2.0cm and no palpable follicles. Her uterus

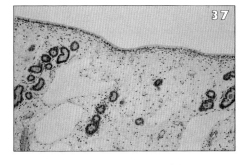

feels small and both uterus and cervix are flaccid. An endometrial biopsy is collected (37). What would you suspect is wrong with this mare?

35–37: Answers

35 i. The precise pathophysiological mechanisms of acute laminitis are uncertain and may be variable between different cases. Potential causes of acute laminar degeneration include vasoconstriction of vessels in the digit, microthrombosis, perivascular oedema, arterio-venous shunting of blood, and a compartmental syndrome due to increased interstitial fluid and pressure. The acute laminar degeneration that results is most marked along the dorsal aspect of the hoof. The distal phalanx detaches from the hoof capsule, especially at the toe, and the weight of the horse transmitted down the bony column and the pull of the deep flexor tendon result in distal movement or rotation of the bone within the hoof. Pressure from the tip of the distal phalanx can result in necrosis of the solar corium and subsequent penetration of the tip of bone through the sole.

ii. The Obel grades are a measure of the degree of pain shown by affected horses. They are also useful prognostically.
- Obel grade 1: the horse lifts its feet incessantly, often at intervals of but a few seconds.
- Obel grade 2: the horse moves willingly at a walking pace, but the gait is characteristic for laminitis. A forefoot may be lifted without difficulty.
- Obel grade 3: the horse vigorously resists attempts to lift a forefoot, and moves reluctantly.
- Obel grade 4: the horse must be forced to move, and may be recumbent.

iii. Subacute laminitis is a mild form of acute disease, which usually resolves quickly and is not associated with any movement of the distal phalanx within the hoof.

Acute laminitis is a more severe clinical disease, often associated with movement of the distal phalanx.

Refractory laminitis refers to acute laminitis that does not improve or respond to treatment, or responds only minimally within 7–10 days. This form of laminitis is often associated with severe laminar degeneration and carries a poor prognosis for recovery.

36 i. Lice, psoroptic mange, *Oxyuris* infection.
ii. Lice.

37 Normal ovarian dimensions are 3 x 3 x 5cm. They can be considerably larger than this when there are multiple medium to large follicles present. The presence of small ovaries can indicate ovarian atrophy in mares which have had debilitating diseases, are malnourished or have been treated with anabolic steroids. In these mares the ovaries were once normal in size and function but have now shrunk in size. Alternatively, they can have ovarian hypoplasia in which the small ovarian size is congenital. Approximately 50% of mares with primary infertility and ovarian hypoplasia have detectable chromosomal abnormalities. In this case the flaccidity of the cervix and uterus, the small ovaries and the endometrial biopsy which shows very few endometrial glands, along with the normal external genitalia, are highly suggestive of 63X gonadal dysgenesis.

38 & 39: Questions

38 A 14-year-old Thoroughbred gelding was presented for swelling of the proximal forelimbs and pectoral region (**38**). Auscultation of the heart was normal as was its rate and rhythm. No murmurs were heard. When auscultating the lung fields, normal breath sounds were heard dorsally while there was an absence of breath sounds ventrally.

i. What is likely to be responsible for the absence of lung sounds?
ii. What are two differential diagnoses?
iii. Name two diagnostic tests to help you differentiate between your two main differentials.

39 A 23-year-old mixed breed horse (**39**) is examined because of failure to shed its winter coat, chronic coughing and, more recently, partial anorexia and loss of weight. A serum chemistry had the following abnormalities: hyperglycaemia (9.2mmol/l; 166mg/dl) and increased liver enzyme activity (AST, 560u/l; GGT, 167u/l). Clinical examination reveals

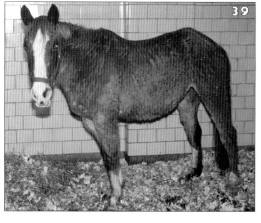

widespread wheezes during thoracic auscultation. The temperature and heart rate are normal but the respiratory rate is elevated (20bpm) and there appears to be an increase in expiratory effort.

i. What diseases are most probable from the clinical examination of this patient?
ii. What further diagnostic tests should be performed to help confirm your diagnosis?
iii. How might the elevation in liver enzymes be related to one of the physically apparent disorders?

38 & 39: Answers

38 i. Pleural effusion.
ii. The majority of all pleural effusions reported in horses are secondary to bacterial pleuropneumonia or thoracic neoplasia. Pleural effusion can occur in any horse with pleuropneumonia. It is not restricted to those horses with severe pleuropneumonia. Lymphosarcoma is the most common neoplasm of the thorax of the horse. Pleural effusion has been reported with other thoracic neoplasms including adenocarcinoma, squamous cell carcinoma, haemangiosarcoma and pleural mesothelioma.
iii. Thoracic ultrasonography and cytological examination of the pleural effusion. Thoracic ultrasonography enables the clinician to characterize the fluid and evaluate the underlying lung. The detection of areas of pulmonary consolidation, necrosis or abscesses and the appearance of a cellular effusion with possible adhesions are suggestive of an infectious process rather than neoplasia. The detection of large volumes of acellular fluid with the pulmonary involvement confined to atelectasis is more likely associated with thoracic neoplasia. Neoplastic masses may be detected within the lung parenchyma or the mediastinal lymph nodes of the horse. Analysis of the pleural fluid will further help differentiate between a septic or neoplastic effusion. When septic effusion exists, total protein and total nucleated cell counts are usually high. The predominant cells are neutrophils and degenerative changes are common. Effusions secondary to neoplasia are less cellular while protein content may remain high. Neoplastic cells may exfoliate and allow for confirmation of the thoracic neoplasia.

39 i. Pituitary adenoma (hirsutism) and chronic obstructive pulmonary disease (COPD).
ii. The age of the horse, clinical signs and hyperglycaemia in combination are virtually diagnostic of pituitary adenoma. Further confirmatory tests might include: thyrotropin releasing factor (TRF) stimulation test, plasma insulin concentration, plasma ACTH concentration (the sample must be promptly frozen and remain frozen until the assay is performed for ACTH measurement). There are many other tests that could be performed but none appear to be as sensitive and specific as the TRF response or ACTH measurement or as easy as the plasma insulin (a single sample is all that is required). A tracheal wash or bronchoalveolar lavage would be indicated to help confirm the diagnosis of COPD.
iii. An ultrasound examination of the liver would be indicated. In this case it did not look abnormal. A liver biopsy might also be indicated since the cause of the elevated liver enzyme activity is not proven. In this case the microscopic diagnosis was hepatic lipidosis. This is likely to be related to the pituitary adenoma and the resulting increase in lipolysis.

40 & 41: Questions

40 What condition is illustrated here (**40a**)? Describe the factors that will influence your evaluation of the case and suggest how the case may be managed.

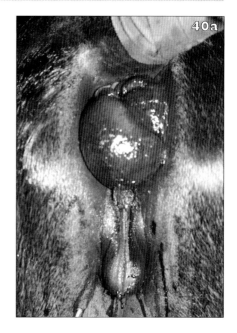

41 This nine-year-old Thoroughbred (**41**) developed an *Aspergillus* spp. fungal keratitis during winter. Aside from the eye problem and a chronic, phenylbutazone-responsive lameness, the horse is healthy. The eye treatment consists of topical clotrimazole and oral phenylbutazone. After five days of treatment, improvement in the eye is noted but the horse suddenly

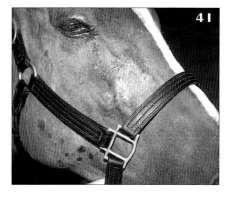

develops a tender facial dermatitis. The skin in the region is very friable and erosions or ulcers could be created with digital pressure. The surface of one of the ulcerated areas was scrubbed, the skin was squeezed and the resultant serosanguinous exudate was smeared and examined cytologically. No fungal organisms or bacteria were seen and the predominant cell type was the eosinophil. What is the most likely diagnosis?

40 & 41: Answers

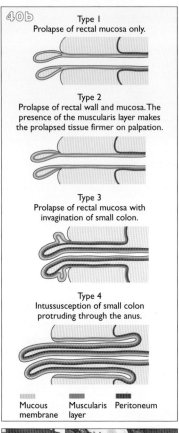

40b

Type 1
Prolapse of rectal mucosa only.

Type 2
Prolapse of rectal wall and mucosa. The presence of the muscularis layer makes the prolapsed tissue firmer on palpation.

Type 3
Prolapse of rectal mucosa with invagination of small colon.

Type 4
Intussusception of small colon protruding through the anus.

Mucous membrane | Muscularis layer | Peritoneum

40c

40 The illustration is of a type 1 rectal prolapse in a yearling which had recently been turned out onto lush pasture. Evaluation of rectal prolapse cases involves predicting what tissues are involved and the degree of damage that has occurred to these tissues (**40b**). Type 1 prolapse, as in the case illustrated, involves just the rectal mucosa protruding from the anus. Type 2 prolapse involves the submucosa of the rectum as well and can be differentiated by the firmer texture of the prolapse on palpation. Most cases of types 1 and 2 can be treated medically and carry a good prognosis, unless the prolapse has gone untreated for many hours or days and extreme mucosal dessication has occurred. Prolapses of types 3 and 4 involve the exposure of much more tissue and are usually candidates for submucosal resection, or resection and anastomosis of devitalized colon. Type 4 prolapse is more accurately described as colonic intussusception and the prolapsed intestine is usually ischaemic due to tearing of colonic blood vessels (**40c**). The prognosis for types 3 and 4 is therefore more guarded.

Management of types 1 and 2 rectal prolapse is aimed at abolishing straining and reducing mucosal inflammation and oedema. Tenesmus and anal or rectal sensation can be abolished by epidural anaesthesia. Once desensitized, the exposed mucosa can be washed, lubricated with a water-soluble lubricant, massaged to reduce oedema and pushed gently back into the pelvic canal. Anti-inflammatory drugs such as flunixin meglumine are useful to reduce inflammation and pain once the epidural has worn off. Some clinicians use a purse-string suture in the anus for 12–24 hours to prevent recurrence of the prolapse.

41 The most likely cause is a drug reaction. The eosinophilic cytology, the absence of visible infectious agents and the friable skin strongly suggest an immunological disorder.

42–44: Questions

42 The Thoroughbred mare described in **37** presents with a history of failure to show oestrus behaviour. You suspect a chromosomal abnormality.
i. How do you confirm a chromosomal abnormality?
ii. What is the most common chromosomal abnormality to present with this history?

43 Shown (**43**) are the chest wall and lungs of a premature neonatal foal. Ribs 3–7 on the right side are fractured along a fairly straight line, with their fracture tips pointing medially. The lungs are slightly expanded but still show congestion, and both the lung and the heart show the fatal puncture wounds caused by the tips of the fractured ribs. What is the probable cause of these fractures?

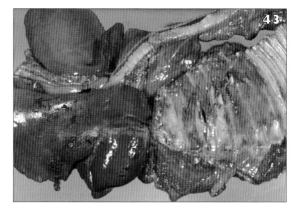

44 The neonatal foal shown (**44**) is being fed milk via an indwelling nasogastric tube.
i. What clinical conditions might require the use of substituted enteral feeding in the foal?
ii. What are the approximate daily milk needs in a foal?
iii. What alternatives exist to supply nutrition to an orphaned foal?
iv. What is the formula for total parenteral nutrition (TPN)?
v. What is the most common complication in a neonatal foal which has recently become capable of nursing unassisted?

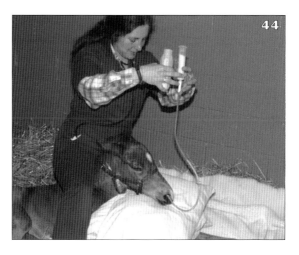

42–44: Answers

42 i. Confirmation of this condition requires karyotyping. A karyotype can be obtained from any actively dividing cell but blood lymphocytes are normally used. Blood samples should be collected into heparin or acid citrate dextrose (ACD) and sent by rapid transit to a reputable laboratory.
ii. 63X gonadal dysgenesis is the most commonly encountered karyotypic abnormality in mares. It is equivalent to Turner's syndrome in women. The condition is sporadic and is thought to develop from nondysjunction during meiosis. Approximately 15% of 63X mares have a second cell line of 63X/64XX or 63X/64XY. 63X mares tend to be smaller than the breed average and are often small and weak at birth. They should be considered to be sterile.

43 Fractures of one or two ribs, or possibly more, can be caused by being stepped on by the mare or another animal. However, when multiple fractures are seen in line, it is most likely the result of parturition occurring when one leg is held back along the chest wall during passage of the foal through the birth canal.

44 i. An orphaned foal, a foal which lacks a suckle response, a foal born to an agalactic or partially agalactic mare, colostrum administration, neonatal isoerythrolysis or a physical inability to nurse properly (eg. wry nose, cleft palate, oesophageal stricture or diverticulum, pharyngeal paralysis).
ii. Approximately 20% of the foal's bodyweight per day, divided into 1–2 hourly feeds.
iii. Milk replacers in liquid or pelleted forms, or goat's milk. In certain clinical situations the use of TPN may be indicated. Some foals with inappropriate suckle responses can be taught to drink from a pan.
iv. 1:1 ratio of amino acids (usually 8.5% solution) with 50% dextrose. Isotonic lipids may be added to the TPN for foals with more severe cachexia. Administration is usually continuous at a rate of approximately 0.6–1.0ml per minute (or 1 drop every 4–6 seconds).
v. Aspiration pneumonia.

45–47: Questions

45 i. How would you treat a horse with lactation tetany (hypocalcaemia)?
ii. What potential complication may develop from bicarbonate treatment of an acidotic, hypocalcaemic horse?

46 Shown (**46**) is the foot of a horse affected by laminitis that has been shod with a heart-bar shoe.
i. What is the rationale for the use of this shoe?
ii. In an emergency situation, how else can you provide frog support in horses with acute laminitis?
iii. What complications can be associated with the use of a non-adjustable heart-bar shoe?

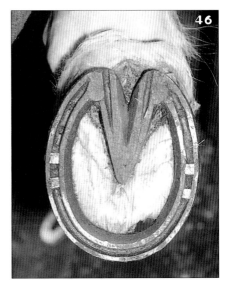

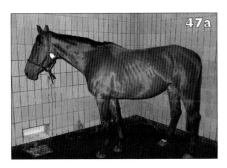

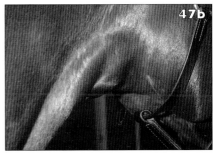

47 You are asked to examine a 14-year-old Warmblood mare that has developed weight loss over a period of six months (**47a**). On palpation, the right thyroid gland feels enlarged (5 x 3cm) (**47b**).
i. Is hyperthyroidism a likely diagnosis?
ii. How would you further examine the mass?
iii. Can it be a malignant mass?
iv. How would you treat this horse?

45–47: Answers

45 i. The disorder is easily treated with intravenous administration of calcium. Equine tetany can be treated by administering 55ml/100kg bodyweight of a standard bovine calcium solution (approximately 11g calcium gluconate/500ml) slowly intravenously whilst auscultating the heart. In addition, it is important to use feeds with a good calcium:phosphorus ratio, and to avoid bran and barley.
ii. The majority of blood calcium is protein bound. Ionized calcium is known to be the biologically active component. Alkalosis increases the protein-bound fraction of calcium, whereas acidosis increases the concentration of ionized calcium. Alkalosis should be prevented as it exaggerates the hypocalcaemic state. In cases of hypoalbuminaemia, signs of hypocalcaemia can be absent due to a normal plasma concentration of ionized calcium. The role of magnesium in clinical cases of hypocalcaemia in horses remains unclear.

46 i. The heart-bar shoe provides frog support, which helps to stabilize and support the distal phalanx. This support decreases the strain on the compromised laminae, which decreases or prevents further distal phalangeal rotation and sinking. Frog support also reduces pain, and improves digital circulation by relieving the pressure exerted by the distal phalanx on the solar plexus, circumflex vessels and vessels of the dorsal coronary corium.
ii. Short-term frog support can be provided with pads constructed from rolls of gauze or commercially manufactured pads taped to the feet.
iii. Difficulty may be experienced in providing the optimal amount of pressure that the non-adjustable heart-bar shoe places on the frog. Too much pressure can cause increased pain and lameness; too little pressure fails to provide support to the distal phalanx. It is difficult to maintain constant pressure on the frog as the heel grows out. Frequent removal and re-application can damage the hoof wall. Abscess formation and granulation under the frog can sometimes occur.

47 i. No. Hyperthyroidism has not been reported in the horse. Asymmetrical hyperplasia of one of the thyroid glands is a common finding in an aged horse. The lesion is classified as an endocrine inactive microfollicular adenoma. Thyroid gland function is usually normal.
ii. Although usually not necessary, a fine needle aspirate or ultrasonography can be used for evaluation of the enlarged gland.
iii. Yes. However, reports of thyroid carcinoma and parafollicular (C-cell) tumours in the horse are rather rare. A thyroid follicular carcinoma with metastasis (and a pituitary pars intermedia adenoma) has been reported in a 15-year-old horse.
iv. In general, treatment is not necessary, unless the palpably enlarged thyroid gland causes obstruction or compression of adjacent structures such as the trachea.

48–50: Questions

48 A six-year-old Thoroughbred mare is presented for artificial insemination with frozen semen from a Warmblood stallion. She is in oestrus and has two large pre-ovulatory follicles on her right ovary – one is 35mm and the other 38mm in diameter.
i. Are you going to carry on and inseminate the mare at this oestrus?
ii. If you decide to inseminate her, how are you going to manage timing of insemination to obtain optimum pregnancy rates?

49 Shown (**49**) is an eight-year-old pony gelding with weight loss (approximately 45kg), inappetence and non-painful pitting oedema of the ventral abdomen. The heart sounds normal, the heart rate is 40bpm and the jugular veins are not distended. The pony is azotaemic (creatinine 194.5µmol/l; 2.2mg/dl) and the urinalysis reveals specific gravity of 1.014 and proteinuria (urine

protein:creatinine ratio 3:1). The PCV of the pony is 35% and the plasma protein is 45g/l.
i. What is the most likely diagnosis in this pony?
ii. What is the presumed pathological mechanism of the disease?

50 Diarrhoea has occurred in the majority of foals on an intensively stocked Welsh Mountain pony stud farm (**50**). Typically, the foals develop diarrhoea at between two and four weeks of age but most do not develop any serious systemic illness and none have died. The stud owner is concerned by the high prevalence of diarrhoea in the foals and also that in many individuals the signs persist for up to 10 days. Of diarrhoeic faeces samples from the affected foals submitted to a diagnostic laboratory, 20% are positive for *Eimeria leuckartii* infection. What is the significance of the laboratory findings?

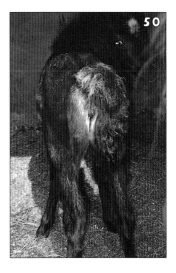

48–50: Answers

48 i. Mares prone to double ovulations tend to repeat multiple ovulations throughout the summer, so there is little point in delaying insemination until the next oestrus. Mares which ovulate two follicles have higher pregnancy rates than mares ovulating one follicle. Therefore, so long as the pregnancy is managed and monitored correctly, it is advantageous to inseminate mares with two pre-ovulatory follicles.

ii. When using frozen semen, mares should be inseminated within six hours of ovulation. There are three ways of achieving this:
- Scan the mare's ovaries every six hours until ovulation is detected. Inseminate at the first examination after ovulation is detected. This is very labour intensive, but ensures that only one insemination dose is used.
- Scan the mare's ovaries every 12 hours. Inseminate at the examination prior to predicted ovulation. Scan 12 hours later when, hopefully, ovulation will have taken place, and re-inseminate. The problem with this method is that, at the very least, two insemination doses will be used. If prediction of ovulation is inaccurate, several doses could be used.
- Administration of a single intravenous dose of human chorionic gonadotrophin (hCG) to mares on the first day that a follicle of 35mm in diameter is detected on her ovaries will induce ovulation in the vast majority of mares between 36 and 48 hours later. Therefore, this drug can be used to help control when ovulation will occur, limiting the number of inseminations to two. The mare can be inseminated once, 36 hours after administration of hCG.

49 i. The most likely diagnosis is glomerulonephritis with renal failure.

ii. The presumed pathological mechanism of glomerulonephritis is persistent immune complex deposition or *in situ* formation in the glomerulus. In the horse the deposits are most severe within the glomerular capillary walls. These deposits cause activation of soluble and cellular mediators within the kidney, causing protein loss in the urine, decreased glomerular filtration and renal failure.

50 Almost certainly the identification of coccidian oocysts in the diarrhoeic faeces samples is an incidental finding. In fact, in surveys it was reported that 40–60% of apparently healthy foals were shedding *E. leuckartii* oocysts in their faeces. Moreover, in experimental infection of ponies with *E. leuckartii* no clinical disease was detected. In this group of pony foals affected by non-fatal diarrhoea at less than one month of age the most probable cause is rotavirus infection which can be investigated by faecal ELISA methods.

51 & 52: Questions

51 This radiograph (51) was obtained from an eight-year-old Arabian horse fed a diet comprising principally alfalfa hay .
i. What is your diagnosis?
ii. What treatment would you recommend?
iii. What preventive measures might you suggest?

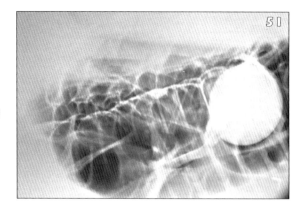

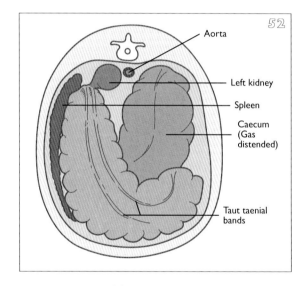

52 You are asked to examine a seven-year-old Warmblood gelding who has been showing signs of mild abdominal pain for two hours. Physical examination is unremarkable; his mucous membrane colour is pink, he has a heart rate of 36bpm and gut sounds are present, although quieter than might normally be expected. Gastric reflux yields approximately two litres of fluid. Findings on rectal examination are represented diagrammatically (52). Does this description allow you to make a diagnosis?

51 & 52: Answers

51 i. Enterolithiasis.
ii. Surgical removal of the enterolith.
iii. Consider altering the diet to one that contains an increased proportion of concentrate and a grass hay to promote a more acidic environment in the colon, because many enteroliths are composed of struvite which will precipitate in an alkaline environment. Evidence exists that feeding a cup of vinegar daily may acidify the colon and thereby decrease the probability of enterolith formation. If a nidus such as a wire can be identified, efforts should be made to minimize environmental contamination with potential nidi (eg. from hay baling wire or worn fencing).

52 Physical examination of this horse revealed no signs of cardiovascular compromise. The most significant findings are from rectal examination which reveal displacement of the left colon dorsally, with taut taenial bands converging towards the nephrosplenic space in the left dorsal quadrant of the abdomen. Gas distension of the colon is also present. This combination of findings is characteristic of nephrosplenic entrapment (NSE) or left dorsal displacement of the large colon. In some cases, gas distension may be so severe that it is not possible to palpate beyond the caudal abdomen. Caudomedial displacement of the spleen and impacted food material in the obstructed colon are other findings sometimes recognized with NSE. The amount of fluid retrieved by nasogastric intubation is of questionable significance; it is, however, quite compatible with NSE. Gastric distension can be caused by pressure from distension of an obstructed left ventral colon on the duodenum which is anatomically closely related to the nephrosplenic space in the left dorsal quadrant of the abdomen.

53–55: Questions

53 A 10-year-old Thoroughbred horse was pulled up during a point-to-point when the jockey realized it was suddenly staggering and very weak. Immediately afterwards, its heart rate was very rapid and did not return to normal for over an hour. Once it became normal, a coarse grade

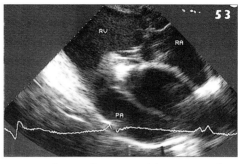

4/6 pan-systolic band-shaped murmur was detected over a wide area on the right hemithorax and the jugular pulsation extended to the top of the neck.
i. What does the echocardiogram (**53**) show? (RA = right atrium, RV = right ventricle, PA = pulmonary artery.)
ii. What causes this condition?
iii. What is the prognosis?

54 You examine a four-month-old foal with bilateral swelling behind the ramus of the mandible (**54**). In the past few weeks a number of the young horses on this large farm have had bilateral purulent nasal discharge, fever and submandibular lymphadenopathy. Only this foal (**54**) has the large swelling just caudal to the ramus of the mandible.

i. What is the likely diagnosis?
ii. What is the aetiology of this condition?
iii. How would you treat this foal?

55 An eight-year-old Quarterhorse (**55**) was examined for acute onset of head tilt to the left, muzzle pulled to the right and difficulty maintaining balance, which rapidly led to recumbency and torticollis. The horse was apprehensive but appeared bright and alert, eating even while recumbent.
i. What is the most likely neuroanatomic location of the disease?

ii. What diagnostic procedures would you like to perform?

53–55: Answers

53 i. A portion of the tricuspid valve everting into the right atrium, indicating rupture of a chorda tendinae of the tricuspid valve.
ii. Rupture of chordae tendinae can occur spontaneously or be preceded by bacterial endocarditis or degenerative valvular disease.
iii. The long-term prognosis is poor. However, the horse may remain comfortable for months or possibly even years provided he is permanently retired from work. Eventually, signs of right-sided heart failure may occur, including jugular venous distension, pleural effusion and ventral oedema. Rupture of chordae tendinae of the mitral valve is likely to cause more rapid progression of clinical signs since the regurgitant fraction enters the pulmonary circulation, causing pulmonary hypertension, pulmonary oedema and eventually right ventricular dysfunction.

54 i. Guttural pouch empyema.
ii. Guttural pouch empyema is frequently seen following infection by *Streptococcus equi* subspecies *equi*. While many horses with strangles develop guttural pouch empyema, the condition is often self-limiting and rarely results in severe distension of the guttural pouches.
iii. Guttural pouch empyema may respond to daily irrigation with physiological saline solutions. The purpose of topical infusions is to dislodge and remove debris from the guttural pouch. Infusions should be repeated once or twice daily until the infection has resolved. Topical antibiotics are rarely effective in guttural pouch empyema because they are unable to penetrate tissues or kill organisms within the brief contact time achieved, and many are inactivated by the products of inflammation. Systemic treatment of guttural pouch empyema is rarely indicated unless there is evidence that the infection is spreading and involving other tissues. If the response to irrigation is poor or if the purulent material becomes inspissated, surgical drainage of the guttural pouch should be considered. An hyovertebrotomy incision combined with ventral drainage through Viborg's triangle or a modified Whitehouse incision are the approaches of choice.

55 i. The horse is most likely affected by peripheral vestibular syndrome. The facial paresis/paralysis, head tilt and balance problem all point to a facial and vestibular nerve dysfunction. The mental attitude of the horse and absence of other cranial nerve deficits suggest that the problem is peripheral rather than central.
ii. Lateral oblique or, ideally, dorsoventral radiographs of the tympanic bulla area may reveal proliferation of the bone in this area. If the stylohyoid bone is affected, a bulging of the bone may be noted on endoscopic examination of the left guttural pouch. In this case there was no bony proliferation but an exudate was noted draining from the left ear. A culture from the ear revealed a heavy growth of *Staphylococcus aureus*.

56–58: Questions

56 Shown (56) is the eye belonging to a 15-year-old Thoroughbred mare that has suffered recurrent bouts of uveitis for over six years. Although the owners have worked diligently to treat these episodes at your direction, this eye has become chronically painful and has not responded to topical corticosteroids, topical 1% atropine and oral nonsteroidal anti-inflammatory drugs. You examine the eye and find this dramatic corneal lesion that does, in fact, retain some fluorescein stain in small punctate areas. What is this, and how should it be treated?

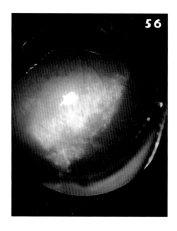

57 A six-year-old Thoroughbred mare is presented for artificial insemination with frozen semen from a Warmblood stallion. She is in oestrus and has two large pre-ovulatory follicles on her right ovary – one is 35mm and the other 38mm in diameter. When are you going to scan her for pregnancy (see **48**)?

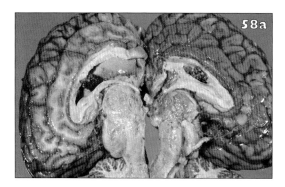
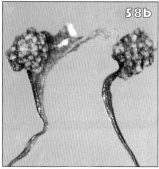

58 When one studies the hemisected brain of this nine-year-old, essentially clinically normal horse, the dark triangular tissue mass (58a) in both lateral ventricles contains a smaller clump of 1mm, pale yellow globules (58b).
i. What are these structures called (both the yellow globules and the larger structures they are in)?
ii. What is their clinical significance at this stage or later?

56–58: Answers

56 Calcific keratopathy. This is a degenerative condition of the central cornea that occurs in some horses that have suffered from multiple uveitis episodes. The lesion is created by calcium deposition at the basement membrane level and superficial stroma. Keratectomy is necessary to treat this condition.

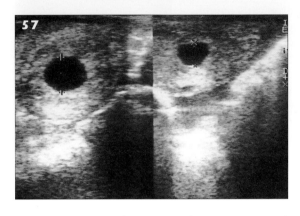

57 It is important that at the examination of this mare, when she is first presented for insemination, a record is kept of the size and distribution of any uterine cysts. Cysts can be the same size and shape as embryonic vesicles. An ultrasound scan of an embryo (*left*) and cyst (*right*) is shown (**57**). If, at pregnancy diagnosis, there is any doubt as to whether embryos or cysts are being scanned, re-examine 3–4 days later. Structures that do not change in size, shape or position, and do not develop an embryo with a visible heartbeat, are likely to be cysts.

Mares can be scanned for twins from 12 days after ovulation. Even in mares in which one ovulation is detected, it is not uncommon for the mare to go out of oestrus and have a second asynchronous ovulation several days later. Sperms from a fertile stallion can retain their fertilizing capacity for several days and therefore it is possible that late ovulation could result in a conceptus which is several days younger than the first conceptus. This latter conceptus could be missed at an early pregnancy check if it is less than 11–12 days old at the time. It is important, therefore, that a repeat scan be carried out to check for twins before 30 days, even if only a single ovulation was detected at oestrus.

58 i. The triangular masses are the normal choroid plexi of the lateral ventricles. The small yellowish round globules are cholesterolaemic granulomas (previously known as cholesteatoma).
ii. At this small size they are not clinically important, and even if they are a great deal larger, they are often found at necropsy examination as an incidental finding. Some, however, do reach a size which causes clinical signs of central nervous system dysfunction.

59–61: Questions

59 A four-year-old miniature donkey is examined for depression and anorexia. The donkey had oesophageal choke two days earlier. The choke had been relieved easily by passing a nasogastric tube. Antibiotics were administered (penicillin and trimethoprim/sulfa) and feed was withheld for 48 hours. On the day feeding was to be resumed the donkey would not eat and was very depressed. The only other abnormal examination findings were mild pitting oedema on the ventral abdomen, tachycardia (92bpm) and tachypnoea (36bpm). Blood was collected for CBC and serum chemistries. The plasma and serum are shown in **59**.
i. What is the diagnosis in this case?
ii. What treatment should be offered?

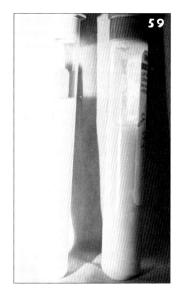

60 The 'curbed' appearance of the hocks of the premature foal shown in **60** is an indication of what clinical problem?

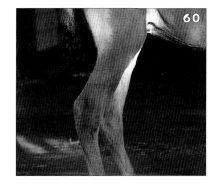

61 A group of 6–12-month-old foals are affected by a cough, mucopurulent nasal discharge, mild pyrexia and slight enlargement of the submandibular lymph nodes (**61**).
i. What is the most likely cause of these animals' disease?
ii. How would you achieve a specific diagnosis?
iii. What treatments would you use?

59–61: Answers

59 i. The donkey has hyperlipaemia syndrome.
ii. The most important treatment is nutritional support (enteral and/or parenteral feeding) and polyionic fluids. Heparin and insulin are used in some cases but the efficacy of these treatments is not proven.

60 The appearance of these hocks is usually the result of crushed tarsal bones secondary to being turned out for exercise. The immaturity of these bones is also present in the carpal bones of premature foals. Premature foals should be kept stall bound until bone maturity is more complete (approximately 3–4 weeks post foaling). Sequential radiographs are recommended. If 'crushing' has occurred, potential athleticism is not consistently compromised since only the distal row of bones is usually affected.

61 i. These foals are most probably affected by a viral infection (eg. equine herpesvirus 1 or 4, equine influenza). There is probably, also, secondary bacterial infection.
ii. The diagnosis of viral respiratory disease is often presumptive, based on clinical signs and epidemiology. However, in some situations a definitive diagnosis of the specific virus may be required. The three most commonly used methods of diagnosis are virus isolation, serology and detection of viral particles/antigens by immunological techniques.

Virus isolation can be attempted from nasopharyngeal swabs. Large gauze swabs are most reliable. These should be passed via the nose to the nasopharynx, where they are left for approximately two minutes. On removal, the swab is placed into suitable virus transport medium and transported immediately to the laboratory, preferably at 0–4°C. In general, samples obtained during the first 24–48 hours of illness are most likely to provide successful cultures. In a group of horses it may be advisable to sample not only those horses showing clinical signs, but also unaffected in-contact animals, as they may be in the early phase of infection and may be more likely to be shedding virus.

Serological tests can be used to measure antibody levels in the blood to specific viruses. Paired samples taken 10–21 days apart are usually required so that a rising antibody titre can be identified. Acute samples must be collected early in the course of disease so that seroconversion is not masked by a titre that is already rising at the time that the acute (baseline) sample is collected. Serological tests, therefore, provide a retrospective diagnosis only.
iii. No treatment may be necessary. If the foals become systemically ill, or if the illness is prolonged (eg. six weeks or longer), antibiotic therapy may be helpful to control secondary bacterial infections. Non-steroidal anti-inflammatory drugs can be used to treat horses with a high fever, inappetence or stiffness. Immunostimulants have been advocated, but there is little evidence of their efficacy to support their use.

62–64: Questions

62 You are asked to examine a three-year-old Thoroughbred mare with a history of intermittent coughing and exercise intolerance. The cough is said to be most pronounced when the horse is eating. The animal is clinically healthy but during light exercise you hear an abnormal noise during both inspiration and expiration. Endoscopic examination reveals the lesion shown in **62**.

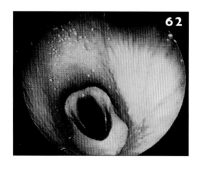

i. What is your diagnosis?
ii. Describe two basic methods for surgical correction.
iii. What postoperative complications may occur?

63 You are asked to examine a three-year-old gelding, purchased 10 days earlier, that has developed a bilateral purulent nasal discharge (**63**), pyrexia (39.5°C), mild cough, depression and inappetence. The submandibular lymph nodes are enlarged and tender. Another two horses in adjoining stables appear to be developing similar signs.
i. What is the likely diagnosis?
ii. What is the aetiology of this condition, and how could you confirm the diagnosis?
iii. How is the disease spread?

64 A 10-year-old Standardbred (**64**) is presented for acute colic, fever (40°C) and fluidy gut sounds on abdominal auscultation. The mare is severely sweating, has tachycardia (90bpm) and tachypnoea, and there appears to be a quick abdominal movement which correlates in number with the heart rate. The mare also has abdominal distension, appears stiff and ataxic, and has trismus of the facial muscles.

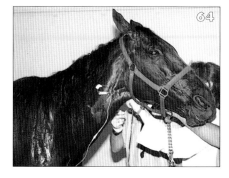

i. What is the unusual abdominal movement?
ii. Which electrolytes would be predictably low in this case?

62–64: Answers

62 i. Epiglottic entrapment. The epiglottis is covered by the entrapping membrane, thereby obscuring the serrated edge and its surface vasculature.
ii. The entrapped epiglottis can be approached through a ventral midline laryngotomy, after which the aryteno-epiglottic tissue is resected. Alternatively, the aryteno-epiglottic fold can be approached through the mouth or nose, after which the fold is divided in the midline using either a hook with a cutting edge on the inside of the hook, or transendoscopically using Nd:YAG laser, or transendoscopically using monopolar electrosurgical cutting equipment. Transendoscopical treatment can be performed in the standing animal.
iii. Dorsal displacement of the soft palate may occur, especially after resection of the aryteno-epiglottic fold. This is most likely to occur if dorsal displacement of the soft palate was observed prior to surgery, or if the epiglottis is hypoplastic or deformed.

63 i. Strangles.
ii. Infection by *Streptococcus equi* subspecies *equi*. The diagnosis is usually based on the characteristic clinical signs, but may be confirmed by culture of the organism from swabs taken from the nasal cavity or from a lymph node abscess. Culture of pus from intact strangles abscesses usually yields a profuse growth of *Str. equi*, but once the abscess has ruptured, secondary infection by other streptococcal species commonly occurs, so isolation of *Str. equi* may become difficult. Likewise, culture of nasal swabs may fail to yield *Str. equi*, especially later in the course of the disease.
iii. *Str. equi* is an obligate parasite of the horse, donkey and mule. The organism is shed in discharges from clinical cases. Transmission is either by direct horse to horse contact (nose to mouth or nose to nose) or by aerosol, or indirectly by contaminated flies and fomites (eg. water buckets, tack, walls, personnel, etc.). The disease is highly contagious, especially under conditions of overcrowding and poor sanitation. Recovered cases of strangles will sometimes shed the organism for a period (usually less than six weeks) after recovery from clinical disease, and such carriers are important in the initiation of new outbreaks of disease.

64 i. The unusual abdominal movement is synchronous diaphragmatic flutter (SDF). This occurs in association with abnormalities in phrenic nerve depolorization and it is activated with each atrial contraction.
ii. Hypocalcaemia is always found with this condition and is the physiological cause of the abnormal phrenic nerve activity. The severe sweating and impending colitis would both result in hypochloraemia. The mare was treated with polyionic fluids and 23g of calcium borogluconate. The bloat and trismus resolved within one hour of the calcium therapy and the SDF disappeared within 12 hours.

65 & 66: Questions

65 A 19-year-old pony gelding is evaluated because of weight loss and hirsutism (**65a**) prior to treatment and (**65b**) 12 months after 0.025mg/kg bromocriptine p/o bid. Blood biochemistry shows hyperglycaemia (18.1mmol/l; 326mg/dl). Urinalysis is normal except for marked glucosuria. Your tentative diagnosis is a pituitary pars intermedia adenoma with secondary diabetes mellitus, which is confirmed by dynamic endocrinological function tests.

i. What is the aetiology of this condition?
ii. How would you treat this horse?
iii. Which plasma electrolyte changes do you expect in the horse after bilateral adrenalectomy, and what is the most important complication of this operation?
iv. Where is cortisol produced and which factors influence it?
v. Is steroid hepatopathy common in equine Cushing's disease?
vi. Which pituitary pathology is associated with a history of fever and nasal discharge prior to acute neurological disease?

66 List the common causes of haemolytic anaemia in the horse.

65 & 66: Answers

65 i. Pituitary-dependent hyperadrenocorticism in the horse arises almost exclusively from an adenoma of the pars intermedia. Equine hyperadrenocorticism associated with either adrenal tumours or ectopic ACTH production has not been reported. The syndrome appears to be the result of an overproduction of POMC-derived peptides in addition to space-occupying effects resulting in dysfunction of the hypothalamus and neurohypophysis, due to the incomplete diaphragma sellae.

ii. Treatment options in equine Cushing's disease include bilateral adrenalectomy and medical therapy. Trans-sphenoidal pituitary microsurgery seems technically impossible in the horse. Bilateral adrenalectomy is unattractive compared to medical therapy. Medical therapy includes two classes of drugs: dopamine agonists and serotonin antagonists (eg. cyproheptadine HCl up to 1.2mg/kg bodyweight p/o sid); as in the pars intermedia, secretion is under tonic inhibitory control by dopaminergic drugs and serotonin antagonists. Dopamine or dopamine agonists (bromocriptine up to 0.03mg/kg bodyweight p/o bid, or pergolide up to 0.011mg/kg bodyweight p/o sid) and/or serotonin antagonists reduce pro-opiolipomelanocortin peptide (including ACTH) secretion from pars intermedia tumour cells. The adrenocorticolytic agent, opDDD, or mitotane is usually ineffective in the horse. In some cases, the elevated plasma glucose concentration is normalized following medical treatment within about 30 days, in combination with shedding of the haircoat.

iii. As the horse cannot retain sodium, chloride and/or water without endogenous mineralocorticoids, pathophysiological changes after bilateral adrenalectomy are decreased serum sodium and chloride, and increased serum potassium and PCV. Also, hypoglycaemia will develop. The most important complication is haemorrhage from the vena cava during dissection of the right adrenal gland, or when it is removed.

iv. Cortisol and corticosterone are the primary glucocorticoids produced by the equine zona fasciculata. The biological half-life of equine cortisol is about two hours. Cortisol concentrations are not affected by breed, age, sex or pregnancy, but are elevated by exercise, fasting and hypoglycaemia, disease and surgery. Since the horse has relatively low concentrations of cortisol-binding globulin, a small increase in total plasma cortisol results in a large rise in bioactive free cortisol. Glucocorticoids are essential to the regulation of carbohydrate and lipid metabolism. Their production is stimulated by ACTH secreted from the anterior pituitary gland. Usually, glucocorticoids strongly inhibit pars distalis ACTH secretion, but do not appear to regulate pars intermedia secretion of POMC peptides.

v. Steroid hepatopathy in horses develops more easily following the administration of triamcinolone acetonide than in association with a pituitary adenoma.

vi. The presence of a pituitary abscess. Treatment with trimethoprim/sulphonamide or third generation cephalosporins is usually futile, because of the advanced stage of disease when neurological signs have developed.

66 (a) Non-infectious causes are neonatal isoerythrolysis, red maple leaf toxicity, onion toxicity, incompatible blood transfusion, phenothiazine toxicity, monensin toxicity, immune-mediated haemolysis.
(b) Infectious causes are equine infectious anaemia, babesiosis, clostridial infections.

67–69: Questions

67 A six-year-old Thoroughbred mare is presented for artificial insemination with frozen semen from a Warmblood stallion. She is in oestrus and has two large pre-ovulatory follicles on her right ovary – one is 35mm and the other 38mm in diameter. How are you going to manage her pregnancy if you diagnose unicornuate twins at 15 days or at 20 days?

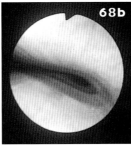
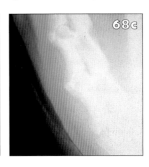

68 The owner of the eight-year-old Miniature horse (**68a**) complained that the horse had been making loud wheezing noises for the past several months. The noise was now beginning to bother her. On physical examination, the respiratory rate was 24bpm and there were loud, stertorous noises which increased as the horse became more excited. When standing by the horse's side, loud inspiratory and expiratory wheezes could be heard. Endoscopic and radiographic appearance of the trachea are illustrated in **68b** and **68c**, respectively.
i. What is this condition?
ii. What is the usual signalment for equine patients with this condition?
iii. What is the pathophysiology of this condition?

69 This five-year-old horse (**69**), just purchased from a farm with a poor worming programme, developed these alopecic, scaly, crusty lesions of its forehead in September. Unless the lesions are scratched or rubbed, they are asymtomatic and have remained static for over two months. No other horse in the barn has developed any skin lesions.
i. List two differential diagnoses.
ii. What is the most likely diagnosis?
iii. What management factor could prevent the appearance of new lesions?

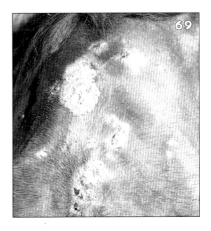

67–69: Answers

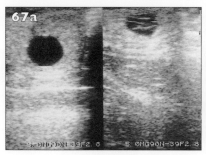

67 Scans for pregnancy performed before 16 days are before the time of embryo fixation, and those performed later are post fixation. The advantage of scanning before 16 days is that the embryos are still mobile. If the embryos are adjacent within one uterine horn at scanning, either the mare can be left for 30 minutes and re-scanned in the hope that the embryos will have moved apart, or the embryos can be moved apart relatively easily by digital manipulation or with the ultrasound probe. Once they are separate, one of the embryos can be pinched between finger and thumb or pushed ventrocaudally against the pelvis with the ultrasound probe. The only disadvantage with this technique is that the younger embryos tend to be more resilient to crushing and it may be necessary to have several attempts before the vesicle bursts. An ultrasound scan of a healthy conceptus (*left*) and crushed twin (*right*) is shown (**67a**). Within seconds this embro had disappeared.

On day 16 the embryo can fix in the same horn or in different horns. If the embryos are fixed in the same horn adjacent to one another, there is a very high rate of natural embryo reduction. 19-day adjacent conceptuses are shown in **67b**. However, because approximately 15% do not reduce naturally, it is worthwhile attempting to separate the embryos with the ultrasound probe (although this is not always possible) and crush one. If this is unsuccessful the mare should be re-examined before day 35 to determine whether natural reduction has occurred and a singleton is present. If the embryos are not adjacent at examination, one should be crushed. The earlier in pregnancy crushing is carried out, the more successful it tends to be. After day 30 it is difficult to perform the procedure and success rates fall.

68 i. Tracheal collapse.
ii. Tracheal collapse is usually seen in mature Miniature horses and Shetland ponies.
iii. Tracheal collapse in the horse is thought to be most likely a congenital defect and usually involves the entire cervical and thoracic trachea. The tracheal rings form an arc with widely separated ends. The dorsal tracheal membrane becomes thin and flaccid. An age-related defect in maintenance of cartilage matrix is suspected to be the underlying problem which results in a lack of rigidity of the cartilaginous rings.

69 i. Dermatophytosis, cutaneous onchocerciasis.
ii. Cutaneous onchocerciasis.
iii. The regular use of ivermectin in worming programmes has largely eradicated this disorder.

70 & 71: Questions

70 A three-year-old gelding is presented one month after castration. The owner's complaint is that the gelding is mutilating its penis. Examination reveals a swollen damaged penis with large wounds and necrotic areas (70), and a warm painful swelling at the base of the penis and the scrotal area. How are you going to treat this case?

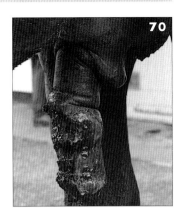

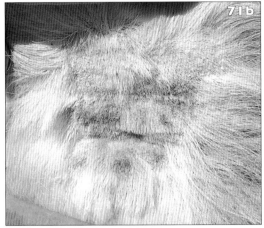

71 The pruritic lesions illustrated in **71a–71c** have been present on the hindlimbs of this eight-year-old Clydesdale gelding for the preceding two months.
i. What is the most likely diagnosis?
ii. How would you treat this condition?

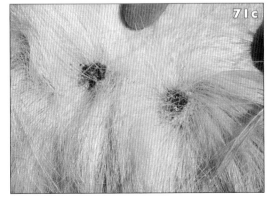

70 & 71: Answers

70 This gelding has post-castration infection with swelling due to haematoma formation and infection, and secondary paraphimosis. Treatment is aimed at surgical drainage of the infected haematoma and penile amputation. Paraphimosis is an inability to retract the penis into the prepuce due to preputial damage (most often traumatic). Paraphimosis should be distinguished from penile paralysis which occurs secondary to reduced retractor penis muscle tone (in debilitating illness, myelitis or spinal injury, or after administration of phenothiazine tranquillizers). Whatever the cause, penile prolapse results in impairment of venous and lymphatic drainage. Within hours there is oedema of the prepuce and penis. Often, swelling is confined initially to the preputial ring which further restricts drainage from the penis and results in penile swelling. If the condition is not treated at this stage, swelling continues to increase and compromises drainage further. Eventually, oedema fluid seeps through the skin, cellulitis can develop and, with additional necrosis, the damage becomes irreversible. In this case the pain caused by the swollen inflamed penis and prepuce caused the horse to self-mutilate.

In cases reported early, before irreversible changes occur, treatment should be started as soon as possible. The primary aim is to support the penis, which will aid in reducing the oedema. If possible the penis should be replaced within the sheath. Numerous devices have been used to support the penis including women's tights, narrow neck plastic bottles and net laundry bags. The supporting device is secured in place by tying tape or tubing from the device over the animal's back – the cranial ties pass over the animal's flanks and the caudal ties between his hindlimbs. The structure must be open or permeable to allow passage of urine. Use of a purse-string suture to retain the penis within the prepuce can lead to abscess and fistula formation.

If it is difficult to replace the penis, oedema can be reduced by massage and ice packs. Emollient creams should be used to keep the penis moist and prevent necrosis. In severe oedema an elastic bandage can be applied to the penis temporarily to reduce the oedema. Once the penis is supported the horse should receive gentle exercise. Systemic treatment with anti-inflammatories, diuretics and antibiotics should be used where appropriate.

In this case the penis was so badly traumatized that the only option was amputation, in association with drainage and flushing of the infected castration site.

71 i. Chorioptic mange. This is a well recognized condition which occurs most commonly on heavily feathered animals, eg. draught breeds. The pruritic lesions typically occur on the lower limbs but on occasions mite infestation can be present on ventral surfaces of the chest or abdomen.
ii. There are no licensed drugs with proven efficacy against chorioptic mange mites. It has been shown that oral ivermectin has limited efficacy in reducing mite numbers and it has been suggested that repeated dosage with the drug at a dose rate of 0.2mg/kg at two week intervals on at least two occasions might produce resolution of clinical signs. From anecdotal accounts of good clinical results following topical application of ivermectin solutions onto the skin lesions at weekly intervals, it would appear that this is the preferred treatment regimen for chorioptic mange. It should be emphasized that this is an 'off-licence' ('extra-label') use of such products.

72–74: Questions

72 The chest radiograph (72a) is of a Thoroughbred horse that has a five-day history of coughing, pyrexia, dyspnoea and inappetence. The smear (72b) is a sample of pleural fluid taken from this horse.
i. What abnormalities are present in the radiograph?
ii. How do you interpret the cytological features of the pleural fluid?
iii. What is the diagnosis?
iv. How would you treat this horse?

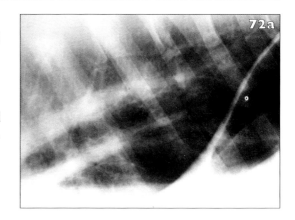

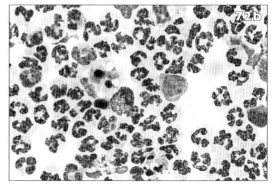

73 *Staphylococcus aureus* can be a pathogen in the horse (see 55). It is most commonly involved in musculoskeletal infections but can occasionally affect other organ systems. What are the best drugs for treating *Staph. aureus* in the horse?

74 This horse was seen to have an acute lid laceration following turn-out in the morning (74). Should you surgically repair this wound, simply allow it to granulate or just snip off the flap of tissue?

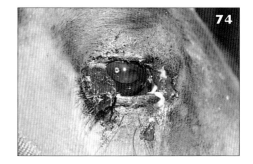

72–74: Answers

72 i. The radiograph shows the dorsocaudal lung field. There is a pleural effusion. There is an absence of aerated lung in the triangular area between the caudal border of the heart and the diaphragm. At the level of the heart base there is an indistinct interface between lung parenchyma above and a homogeneous opacity below, which represents the level of free fluid in which the lungs are partially submerged. The line of the diaphragm is obscured by this homogeneous opacity.
ii. The smear has cytological characteristics of an exudate. The majority of cells are neutrophils, including degenerative neutrophils. The larger cells are macrophages. Bacteria (free or intracellular) may be seen in septic exudates.
iii. Pleuritis (pleurisy). This has probably occurred secondary to pneumonia or lung abscessation.
iv. The most important aspect of treatment is antibiotic therapy. Selection of antibiotics may be made on the basis of culture of bacteria from the pleural fluid.
Additional treatment may include drainage of the pleural effusion. Drainage of pleural fluid allows re-expansion of the lung and is indicated when a large volume of fluid is present or it is frankly purulent in nature. Fluid can be drained using intermittent thoracocentesis or by placing an indwelling chest drain. Pleural lavage may be used in selected cases to aid removal of fibrin and necrotic tissue. In this procedure, two chest drains are inserted, one dorsally and one ventrally, and sterile isotonic solution (eg. saline) is infused into the cavity dorsally and allowed to drain out ventrally. Pleural lavage is contraindicated in the presence of a bronchopleural fistula.

Thoracotomy to permit removal of organized fibrin and debris has been used successfully in some cases that have failed to respond to other treatments. The technique is most beneficial in cases where large, unilateral, localized pockets of thick debris are present after resolution of disease in the contralateral hemithorax. Where adequate expertise and facilities permit, thoracotomy and rib resection can be performed in the standing horse.

73 Proper therapy should be based upon culture and sensitivity tests in each individual case, but as a rule trimethoprim/sulphonamide (trimethoprim/sulfa) is an excellent drug for treating *Staph. aureus* infection in the horse. There are occasional resistant strains which may require cephalosporin or vancomycin therapy but these are uncommon. The horse described in 55 was treated with trimethoprim/sulphonamide and within one week was well enough to be discharged from the hospital.

74 With fine suture material, perform a two-layer lid repair with buried sutures in the lid stroma and exposed sutures in the skin. The conjunctiva does not need to be sutured. Healing by secondary intention or snipping off the flap are contraindicated because they would lead to corneal exposure damage, ulcers or lash or lid abnormalities, such as trichiasis.

75 & 76: Questions

75 A nine-year-old horse is presented for examination prior to purchase. The prospective purchaser wishes to use the horse for International three-day eventing. She knows the horse very well and, in fact, has been riding it in competitions for the current owner for over a year. The horse is currently very successful and shows no performance problems. On auscultation at rest a grade 3/6 holosystolic murmur is detected over the left fifth intercostal space midway between the level of the point of the shoulder and the point of the elbow. The murmur radiates over 2–3 intercostal spaces, caudodorsally. No murmur is audible on the right side. After exercise, the murmur is still audible but the heart rate and recovery are good.
i. What are the major causes for a systole cardiac murmur loudest over the left side?
ii. In this horse, how do the characteristics of the murmur allow the clinician to be more specific about the most likely cause?
iii. How can the clinical diagnosis be confirmed?
iv. Should this horse pass the prepurchase examination?

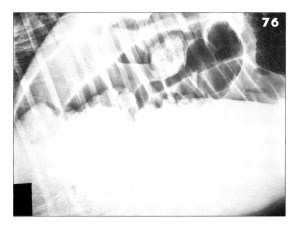

76 You examine a two-year-old Miniature horse gelding for acute abdominal pain. The horse has a two-day history of colic that was initially mild but has progressed slowly to pain that appears moderate to severe. In addition, the owner has noted abdominal distension during the past 24 hours. At admission, borborymi are decreased in intensity, heart rate is 94bpm, respiratory rate is 28 breaths per minute and rectal temperature is 38.2°C. Abdominal radiography is performed (**76**).
i. What is your presumptive diagnosis?
ii. What treatment do you recommend?
iii. What risk factors for this condition do you suggest to the owners?

75 & 76: Answers

75 **i.** The major causes of cardiac murmurs in systole loudest over the left side are mitral valvular regurgitation and physiological murmurs associated with ventricular ejection through the aortic or pulmonic valves.
ii. This murmur has the characteristics of mitral valvular regurgitation, ie:
- Timing – mitral insufficiency is associated with pan- or holosystolic murmurs. Ventricular ejection murmurs are usually mid or early systolic, although they can be holosystolic.
- Point of maximal intensity – the murmur is loudest over the mitral valve area (in contrast, ventricular ejection murmurs are louder over the aortic valve area, left fourth intercostal space at the level of the point of the shoulder or the pulmonic valve, left third intercostal space midway between the level of the point of the shoulder and the point of the elbow).
- Radiation – the murmur radiates widely following the direction of the jet of regurgitation of mitral insufficiency from the left ventricle to the left atrium.
- Quality – the murmur is coarse and band-shaped, typical of mitral valvular regurgitation. Ventricular ejection murmurs are usually softer and have a crescendo–decrescendo shape.
- Intensity – intensity or amplitude is used to grade murmurs. However, both mitral regurgitation and ventricular ejection can cause grade 3/6 murmurs. Therefore, the grade does not help the clinician to differentiate mitral valvular regurgitation from ventricular ejection murmurs. Equally, both murmurs may persist after exercise.

iii. It will be preferable to confirm the diagnosis of mitral valvular regurgitation with echocardiography.
iv. If mitral valvular regurgitation is confirmed, the horse should be judged as not suitable for purchase for competition at an advanced level. Mitral valvular regurgitation is the commonest form of valvular disease associated with poor performance. It is certainly true that horses with mild to moderate mitral valvular regurgitation can perform successfully at this level for a time. However, it is a progressive condition and its presence can predispose the horse to atrial fibrillation. Therefore, despite an accurate history of adequate exercise tolerance, there is a reasonable risk that poor performance will occur in future.

76 **i.** Obstruction of the small colon with a faecalith or perhaps an enterolith; notice that there is also much feed material within the large colon.
ii. Surgical removal via enterotomy is often recommended, although administration of enemas and oral laxatives can be attempted as treatment.
iii. Small breeds of horses may be prone to this condition. The quality of hay may play a role in faecalith formation. Because poor dental condition may prevent horses from adequately grinding ingested hay, it is plausible that dental problems may predispose to this condition.

77–79: Questions

77 The horse described in **52** has a nephrosplenic entrapment (NSE) or left dorsal displacement of the large colon. What are the management options available to you and how would you decide which was appropriate for this case?

78 This 13-year-old Standardbred gelding (**78**) has been diagnosed with a pelvic flexure impaction and is being treated with oral fluids through an indwelling nasogastric tube.
i. What is the volume of the normal equine stomach, and how much volume would you give by nasogastric tube at one time?
ii. What other laxatives, besides water, can be given via a stomach tube to treat large colon impactions?
iii. Once you are confident in your diagnosis of pelvic flexure impaction, is it prudent to administer analgesics during the medical therapy of the disease?

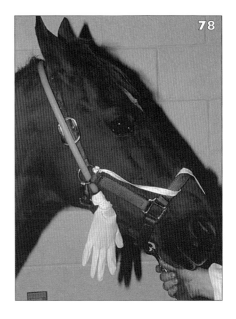

79 A 12-hour-old foal is presented as weak and lethargic. The heart rate is 96bpm, and the mucous membranes are pale and jaundiced (**79**). Laboratory evaluation reveals a haematocrit of 10%, a haemoglobin concentration of 34g/l and a total protein of 52g/l.
i. What is the most likely diagnosis?
ii. What are the three potential clinical entities that could lead to the death of this foal?

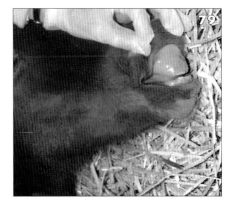

77–79: Answers

77 There are several ways in which a case of this type can be managed:
- *Conservative management.* Some clinicians have achieved good success rates with selected NSE cases. Management involves withholding feed and administering analgesics until the intestinal tract is palpably normal on rectal examination. This can take 24–72 hours. Repeated re-evaluation of the case by rectal examination is necessary to monitor for signs of developing food impaction, tympany or colonic oedema. If any of these problems arise, or if improvement does not occur within 48–72 hours, then exploratory laparotomy is indicated. Case selection is important if conservative treatment is to be used. No horse that has a big colonic impaction, marked colonic tympany or gut wall oedema should be managed in this way.
- *Rolling.* The technique of rolling horses with NSE has been pioneered in Europe, where large numbers of these cases occur in the Warmblood breeds. Various protocols have been described which involve general anaesthesia of the horse (sometimes with barbiturates to induce splenic engorgement) followed by a set routine of suspension from hobbles around the feet, rotation about the horse's long axis, and pummelling of the abdomen to mobilize the entrapped colon. Similar case selection criteria apply as for conservative management.
- *Laparotomy.* Entrapment of the colon in the nephrosplenic space is amenable to correction via a standard ventral midline laparotomy. The advantages of this technique include direct visual assessment of the colon, decompression if tympanitic, and evacuation of colonic impactions via an enterotomy. It should be remembered that (occasionally) NSE can be strangulating and cause colonic ischaemia. Success rates of this surgery are high (90% plus).

78 i. The gastric capacity is 10–20 litres; however, no more than 6–8 litres should be added rapidly to the stomach to prevent pain associated with sudden gastric distension.
ii. Magnesium sulphate (epsom salt), dioctyl sodium succinate (DSS), mineral oil (liquid paraffin), psyllium hydrophilic muciloid.
iii. Yes. Serial palpation per rectum and careful monitoring of systemic signs is critical to ensure that the impaction is responding appropriately to treatment, and to rule out the development of any complicating problems such as large colon displacement. If such monitoring is carried out, analgesics may be administered with safety.

79 i. Neonatal isoerythrolysis.
ii. Three potential clinical entities that could lead to the death of this foal are:
- Hypovolaemic shock secondary to the rapid lysis of red blood cells and their functional masses as vascular expanders.
- Loss of oxygen carrying capacity from the lysed red blood cells.
- The development of kernicterus. The increasingly elevated levels of bilirubin within the bloodstream can cross the immature blood–brain barrier in the foal and result in chemical toxicity to the brain. The brain damage can be permanent and be represented by persistent seizures or maladjustment.

80 & 81: Questions

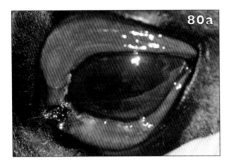

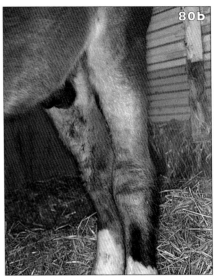

80 You are asked to examine a four-year-old horse from eastern Europe that has developed an illness characterized by pyrexia (41°C), anorexia, depression, conjunctivitis (**80a**), ventral abdominal and preputial oedema, limb oedema (**80b**) and a skin rash.

i. Which viral infection would you be concerned about?
ii. If this horse was a stallion, why might you be concerned about his future use for breeding?
iii. Is it possible to vaccinate against this virus to help control its spread to other horses, and what measures can be taken to prevent its spread by infected stallions?

81 Dehydration (**81a**) is one of the physiological sequelae of endotoxaemia. List the clinical signs associated with dehydration and the laboratory tests available to quantify it.

80 & 81: Answers

80 i. Equine viral arteritis (EVA) (infection by equine arteritis virus).
ii. A carrier state for equine arteritis virus is established in 30–60% of infected stallions. The duration of the carrier state varies from several weeks to lifetime. Carrier stallions shed the virus constantly in the semen. There is no effect on fertility.
iii. A modified live vaccine is safe and effective in stallions and non-pregnant mares. Prevention and control of EVA depend on management practices and selective use of the vaccine, including vaccination of the at-risk stallion population. Stallions should be vaccinated annually at least 28 days before the onset of the next breeding season. Carrier stallions should be kept in isolation and bred only to seropositive mares. Measures should be taken to prevent the spread of equine arteritis virus in fresh or frozen semen used for AI.

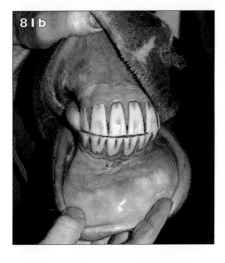

81 Clinical signs associated with dehydration are:
• Poor peripheral pulse quality, sometimes described as weak or 'thready'.
• Increased heart rate when horse at rest.
• Congested mucous membranes progressing to cyanosis as dehydration becomes more severe (**81b**).
• Dry mucous membranes.
• Capillary refill time greater than two seconds.
• Cold extremities.
• Depression, eg. head held low (if not showing signs of pain).
• Increased skin turgor (very subjective).

The single most useful laboratory assessment of dehydration is PCV measurement. This is simple, rapid (result in less than 10 minutes) and with good technique can provide reliable, repeatable results. A refractometer estimation of total solids in the serum (incorrectly referred to as total protein) is often performed at the same time. With both these simple assays, the results should be used to follow trends in individual horses rather than interpreted as absolute values. The repeated monitoring of PCV/total solids should be an integral part of rehydration therapy. Blood gas analysis is far more complex and expensive than PCV estimation, but many of the measured parameters are influenced by dehydration, particularly pH and base excess/anion gap.

82–84: Questions

82 A three-year-old Thoroughbred is examined because of poor performance. On clinical examination the horse is afebrile but has an increased pulse rate (64/minute), jugular distension without pulsation (**82**) and the heart cannot be heard on auscultation.
i. What diagnostic steps should be performed?
ii. How would you manage the case?

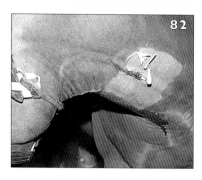

83 The three-year-old Thoroughbred shown in **83** had a 10-day history of rapid weight loss, a reduced appetite (half normal intake), mild dysphagia accompanied by slow chewing and occasional oesophageal spasm during deglutition, rhinitis sicca, muscle tremors and constant sweating over the neck and shoulders. Gut sounds were reduced but still present. A provisional diagnosis of chronic grass sickness was made.

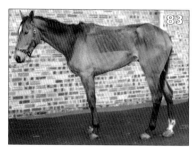

i. How can the diagnosis be confirmed and what problems may arise from the diagnostic procedures?
ii. What is the prognosis?
iii. How would you manage this case?

84 A 21-day-old foal was found down and obtunded in the pasture field. The foal had appeared normal on the previous evening. On clinical examination the foal was severely depressed, tachypnoeic (64bpm), tachycardic (150bpm), hyperthermic (40.2°C), had severe injection of the mucous membranes and appeared slightly icteric.

Abnormal laboratory findings included: leukopenia with a degenerative shift, marked elevation in liver enzyme activity in the serum, hypoglycaemia, severe metabolic acidosis, abnormally high conjugated and unconjugated bilirubin and azotaemia. A liver biopsy was performed (**84**).
i. What is the diagnosis and causative organism?
ii. What is the treatment and prognosis?

82–84: Answers

82 i. An ultrasound examination should be performed on the cardiac area. A large amount of fluid was seen in the pericardial space. This was drained and a sample submitted for both cytology and culture.
ii. The cytology was a non-septic exudate with the predominant cell type being plasma cells and lymphocytes. The horse was treated with dexamethazone, penicillin and gentamicin, and it made a full recovery as a successful racehorse. No bacteria were isolated from the fluid.

83 i. The diagnosis can only be confirmed by histological examination to demonstrate characteristic neuronal degenerative changes in the autonomic ganglia (eg. stellate, anterior cervical or coeliacomesenteric) or the intramural plexuses of the gut, particularly the ileum. This could be carried out at post-mortem examination or ileal biopsies could be collected at laparotomy. In the case described the clinical signs are not too severe and immediate euthanasia is not warranted provided the owner is prepared for the work involved in treatment. However, exploratory laparotomy adversely affects the outcome in chronic grass sickness cases, and with typical clinical signs present, it is questionable whether the animal should be subjected to a laparotomy solely to confirm the diagnosis. The combination of clinical signs described are almost pathognomonic of chronic grass sickness, although clearly each individual sign is not.
ii. The prognosis should always be guarded. However, with appropriate management, approximately 50% of chronic cases recover and 75% of these can be expected to be able to carry out work requiring a high level of fitness afterwards. The fact that the horse is only mildly dysphagic and is thin but not emaciated suggests that it is worth considering treatment. Severe dullness and dysphagia, emaciation and severe rhinitis would have indicated a very poor prognosis with euthanasia being the best option.
iii. The horse should be housed initially but allowed access to pasture for short periods several times daily. High energy, high protein foods which are easily swallowed are preferable, eg. coarse mix, alfalfa nuts or fibre, or crushed oats, possibly soaked in dilute molasses. Succulents, eg. carrots or apples, can be offered to increase interest and palatability. Up to 500ml corn oil can be given daily. Many small, fresh feeds should be offered. The use of rugs which are permeable to water vapour is advisable if sweating is excessive, as they help to keep the animal dry. If gut motility is reduced, cisapride, an indirect cholinergic agent, can be administered orally (0.5–0.8mg/kg bodyweight every eight hours for approximately seven days) to increase gut motility and encourage defecation. If analgesia is required, non-steroidal anti-inflammatory drugs, eg. flunixin meglumine or phenylbutazone, are best as they do not decrease gut motility. If the horse is going to improve, appetite should start to increase within 2–3 weeks from the onset, and weight gain from 3–5 weeks, although complete recovery may take many months.

84 i. The diagnosis is Tyzzer's disease caused by *Bacillus (Clostridium) pilliformis*.
ii. The treatment should include acetate Ringer's solution with dextrose and sodium bicarbonate, antibiotics (eg. intravenous penicillin and an aminoglycoside) and equine plasma. The prognosis is grave.

85–87: Questions

85 A five-year-old horse has died 24 hours after the onset of an acute illness characterized by fever (40.5°C), coughing, dyspnoea and, terminally, a blood-stained frothy nasal discharge (85). The horse has recently (four days ago) travelled from Africa.

i. What viral infection do you suspect?
ii. What is the epidemiology of this disease?
iii. What are the different clinical forms of this disease and their clinical features?

86 You are asked to examine a 10-day-old Coldblood filly. When sucking from the dam, the filly constantly loses milk through both nostrils (86). The foal appears otherwise to be clinically normal.

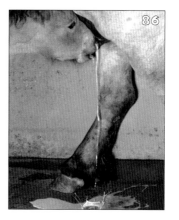

i. What is your tentative diagnosis?
ii. How would you confirm the diagnosis?
iii. What is the differential diagnosis?
iv. What is the prognosis, and how could this foal be treated?

87 Sections of large colon from two different horses are shown (87). The uppermost section is from an adult horse which died suddenly of colic lasting 14 hours. The section of large colon at the bottom, which has several areas of pale serosal surfaces on either side, is from a horse which had a history of weight loss and polyuria of several weeks' duration.

Based on the gross findings, give several possible diagnoses and their suggested pathogenesis.

85–87: Answers

85 i. African horse sickness (AHS).
ii. AHS is endemic in sub-Saharan Africa, but has spread elsewhere on a number of occasions, including Pakistan, the Middle East and Spain. AHS is arthropod-borne and may spread via infected animals or by vector movement. The main vectors are *Culicoides* species of midges. Spread by contaminated needles or surgical instruments is also possible.
iii. Four syndromes are recognized:
- Peracute or pulmonary form: incubation period 3–5 days; fever (up to 40.5°C); congestion of mucous membranes; dyspnoea and tachypnoea; cough; sweating; frothy blood-tinged nasal discharge (terminally); high mortality rate (95%) with death in 1–3 days.
- Subacute or cardiac form: incubation period 7–14 days; fever (up to 40.5°C); congestion of mucous membranes; oedematous swellings of neck, chest, lumbar and pelvic areas; oedema of supraorbital fossa, eyelids, intermandibular space; petechial haemorrhages on tongue and conjunctivae; colic; mortality rate is 50% with death in 4–8 days.
- Acute or mixed form (most common clinical form): incubation period 5–7 days; clinical signs of both pulmonary and cardiac forms; mortality rate 50–95% with death in 3–6 days;
- Horse sickness fever form: occurs in partially immune horses and donkeys; incubation period 5–14 days; low grade fever (up to 39.5°C); conjunctival congestion; mild depression and inappetence; recover in 5–8 days.

86 i. Cleft palate (palatoschisis).
ii. Inspection of the mouth and/or endoscopic examination of the nasopharynx and soft palate.
iii. Excessive milk production by the mare; megaoesophagus; oesophageal stricture; oesophageal obstruction (choke); soft palate dysfunction.
iv. Although there have been a few favourable reports of the surgical repair of cleft palate, the prognosis is poor. Even if palatoplasty is successful, this will not restore athletic capability and the surgery is no more than a salvage procedure. Saving the animal for breeding is unwise, since this condition can be hereditary. If surgical correction is considered as an option, the procedure should be performed as soon as possible, preferably at one day of age. The foal should be treated to prevent aspiration pneumonia for at least five days after surgery.

87 The relatively short duration of colic in an otherwise normal horse, and the marked congestion and oedema of a major portion of the large colon, is suggestive of a large bowel torsion, and this was found at necropsy examination. Most cases occur without a known cause. The extensiveness of the affected bowel tends to rule out vascular compromise from parasite migration or thrombosis. The bottom colon, while similar in appearance, smelled strongly of ammonia, suggesting a diagnosis of uraemic colitis. Toxic and infectious causes should also be considered.

88–90: Questions

88 A 21-year-old Welsh pony mare (non-pregnant) is evaluated because of extremely slow healing of a wound to the left tarsus and laminitis of the right hind foot (88). Distal displacement of the third phalanx of the right hind foot is seen radiographically. The horse damaged its left tarsus four weeks earlier. Blood biochemistry shows a normal plasma glucose concentration without glucosuria.

The serum activity of alkaline phosphatase is slightly increased (427u/l; n = 250–300u/l). On the basis of the delayed wound healing, the laminitis and the increased serum activity of alkaline phosphatase your tentative diagnosis is a pituitary pars intermedia adenoma.
i. Does a pituitary pars intermedia adenoma in the horse most probably arise as a primary pituitary disorder or from loss of the dopaminergic control?
ii. Which tests are available to confirm the diagnosis?
iii. How would you deal with the diagnosis in practice?
iv. Does the presence of an increased activity of thermolabile alkaline phosphatase support the diagnosis?
v. Can hyperadrenocorticism in this horse be ruled out because of normal results of the glucose tolerance test?

89 A three-year-old Throughbred, in racing, develops an acute onset of head tilt to the left, signs of facial paresis on the left, ataxia, dysphagia and moderate depression (89). There is no history or evidence of trauma and radiographs of the skull are normal.
i. What would be the most likely diagnosis based upon the history and clinical findings if the horse is in the US or has been transported to Europe from the US?
ii. What would be the prognosis?

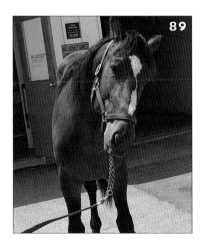

90 List at least five causes of hyperkalaemia in the horse.

88–90: Answers

88 i. The finding that adenomas contain reduced concentrations of dopamine, but not serotonin, supports the latter hypothesis.
ii. The diagnosis of an adenoma of the pars intermedia of the pituitary gland in the horse depends on dynamic endocrinological function tests, such as the dexamethasone suppression test, ACTH stimulation test, insulin/glucose tolerance test and TRH test. In addition, ventrodorsal radiography in combination with a contrast venographic technique or computed tomography are possible. Single samples for basal cortisol concentration are of no diagnostic value, in contrast to basal ACTH concentration and the urinary corticoid:creatinine ratio. In normal horses there is a diurnal rhythm of plasma cortisol, with a peak at 6–9 hours, and the lowest level at 18–21 hours. Affected horses usually fail to exhibit diurnal variation.
iii. The collection of a morning urine sample for the assessment of the corticoid:creatinine ratio might be used as a screening test. A ratio of $>20 \times 10^{-6}$ must be regarded as positive. However, false positive and false negative test results can occur. The overnight dexamethasone suppression test (determining plasma cortisol concentration in blood samples collected immediately before and 19 hours after the intramuscular administration of dexamethasone 40µg/kg bodyweight at 5.00pm) is a reliable clinical test for diagnosing a pituitary pars intermedia adenoma. There is no significant difference in dexamethasone suppression tests initiated at either 9.00am or 9.00pm in normal horses. Horses with pituitary pars intermedia adenoma are highly resistant to glucocorticoid negative feedback inhibition of ACTH secretion and, as a result, plasma cortisol 19 hours after dexamethasone administration will not be lower than 27.6nmol/l. Also, the thyrotropin-releasing hormone response test can be performed by administering 1mg TRH i/v and blood samples collected at hours 0 and 0.5. Plasma cortisol in diseased horses increases significantly after 30 minutes. The ACTH stimulation test can be used by administering 25iu of synthetic $ACTH_{1-24}$ i/v at 9.00am. Blood samples are collected at 0 and 2 hours postinjection. There is an exaggerated cortisol response to exogenous ACTH in pituitary-dependant Cushing's disease (plasma cortisol two hours after ACTH >413nmol/l).
iv. Yes. The alkaline phosphatase activity in cases of equine Cushing's disease is heat sensitive.
v. No. This test can show a normal response in cases of pituitary pars intermedia adenoma.

89 i. Equine protozoal myeloencephalitis (EPM).
ii. Horses affected with EPM and demonstrating acute brainstem signs, especially dysphagia, generally have a more guarded prognosis than those with spinal cord signs alone.

90 Renal failure, uroperitoneum, hyperkalaemic periodic paralysis, metabolic acidosis, strenuous exercise, extensive tissue damage and haemolysis, type 4 renal tubular acidosis.

91–93: Questions

91 You are asked to examine a five-year-old pony mare which has developed a sudden onset of generalized stiffness (**91**). The pony stands with her head extended, and the limbs are placed in a 'sawhorse' stance. The nostrils are flared, the ears erect and stiff, and the lips retracted. The tail head is elevated. There is dysphagia and saliva drools from the mouth.

i. What is your diagnosis?
ii. Which organism is responsible for this disease?
iii. How would you treat this condition?

92 A six-year-old Thoroughbred mare had a bilateral, profuse foamy discharge with some food tingeing out of both nostrils for two days. A nasogastric tube was passed into the stomach with no difficulty. Endoscopic examination showed a large volume of saliva within the pharynx, larynx and trachea, left laryngeal hemiplegia and dorsal displacement of the soft palate. The endoscopic view of the interior of the left guttural pouch is shown (**92**).

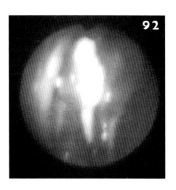

i. What is the most likely diagnosis?
ii. What is the most likely aetiologic agent?
iii. What is the most likely reason for the saliva accumulation?
iv. What is the prognosis?

93 A seven-year-old horse with an asymptomatic, crusted lesion in the bulb region (**93**) had just been purchased. A wire cut was the supposed cause of the lesion. The new owner started to pick at the crust and scrub the area vigorously with iodine. Instead of improving, the lesion started to proliferate and became tender and ulcerated. List three possible causes for the lesion.

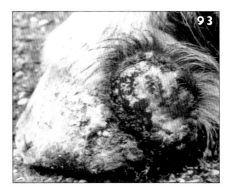

91–93: Answers

91 i. Tetanus.
ii. *Clostridium tetani.*
iii. Treatment aims at providing sedation, muscle relaxation, and suitable bedding, eliminating infection, and administering antitoxin, fluid and nutritional support:
- Sedation and muscle relaxation. Keep the horse in a quiet, dark environment, and undisturbed. Sedate with acetylpromazine, chlorpromazine or promazine at 4–6 hourly intervals. Stronger sedation may be achieved, if necessary, using chloral hydrate or sodium pentobarbitone. Diazepam used alone, or in addition to other sedatives, reduces severe muscle spasms. Other muscle relaxants can be used in combination with sedatives, including guaifenesin (given to effect by slow i/v drip) and methocarbamol. A balance between the amount of sedation and the degree of muscle relaxation is important. An intravenous catheter may be left in place to minimize stimulation when administering drugs. Packing the ears with cotton wool minimizes auditory stimulation.
- Provide adequate bedding for recumbent horses to minimize decubital ulcers. Standing horses should have adequate footing. Consider the use of slings to support horses that tolerate them.
- Eliminate infection by surgical debridement of wounds and parenteral administration of penicillin.
- Antitoxin. Administer a large dose of antitoxin (5000–10000 units) subcutaneously at the onset of clinical signs. Local infiltration of any wound with up to 9000 units of antitoxin has also been suggested. Tetanus toxoid may be administered simultaneously with antitoxin, but at a separate site.
- Fluid and nutritional support. Intravenous fluid and electrolyte therapy may be necessary. Dysphagic horses may be fed through an indwelling nasogastric tube.

92 i. A white effervescent plaque is seen on the caudodorsal region of the medial compartment of the guttural pouch, caudal and medial to the articulation of the stylohyoid bone with the petrous part of the temporal bone. A lesion with this appearance in this region is most likely guttural pouch mycosis.
ii. While a number of fungi have been associated with guttural pouch mycosis, *Aspergillus fumigatus* and *A. nidulans* are frequently isolated.
iii. Dysphagia associated with guttural pouch mycosis is explained by involvement of the pharyngeal branches of the vagus and glossopharyngeal nerves.
iv. Grave; horses that develop dysphagia secondary to guttural pouch mycosis may eventually recover, but this may take at least 6–18 months. Unless the horse is capable of maintaining itself throughout this prolonged recovery period, fluid and caloric needs will need to be provided by nasogastric intubation or a tube placed through an oesophagostomy.

93 Three causes are:
- *Granulation tissue*: due to excessive wound treatment.
- *Granulomatous sarcoid*: conversion to more aggressive form by harsh treatment.
- *Botryomycosis*. This horse had botryomycosis. The aggressive wound handling was probably not responsible for its development.

94–96: Questions

94 Which of the commonly available sterile fluids would you choose to rehydrate an endotoxaemic horse prior to colic surgery? Justify your choice to the anaesthetist.

95 A two-month-old colt foal is presented with a bilateral swelling of the scrotum (95) which has been present since the foal was five days old. The swelling is soft and painless.
i. What is the most likely diagnosis?
ii. How would you treat this foal?

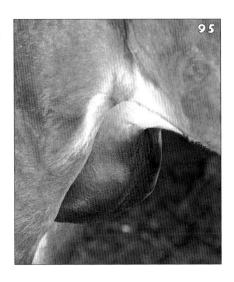

96 This 12-year-old Percheron stallion had a prolonged history of bilateral forelimb lameness that was responsive to intermittent administration of phenylbutazone by the owner. Over the past two weeks he has become anorexic, is losing weight and has developed generalized dependent oedema (96). His heart rate is mildly elevated and temperature and respiratory rate are normal.

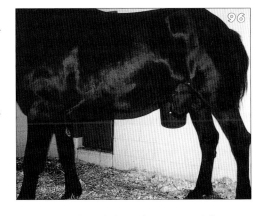

Oral mucous membranes are mildly congested and the sclera are mildly injected.
i. What is your primary differential diagnosis and diagnostic plan related to this disease?
ii. Inflammation of what part of the gastrointestinal tract has been specifically associated with NSAID toxicity in horses?.

94–96: Answers

94 The major problem with fluid therapy in the horse relates to the volume required. Assuming 10% dehydration in a 500kg horse, a volume of 50 litres must be given to make good the deficit. It is useful to consider two different types of fluid which have application in this situation:
- Isotonic fluids (eg. Hartmann's or Ringer's solution, Isolec, 0.9% sodium chloride) when given intravenously will immediately redistribute to the extra-cellular space thus providing only slow improvement in circulating fluid volume. Thus large volumes must be given and this takes time. Time is an important factor in many surgical colic cases and the progression of ischaemic bowel lesions precludes hours of pre-surgical fluid therapy. The advantages of such therapy are that balanced electrolyte solutions will replace electrolytes (in physiologically appropriate proportions) as well as fluid, and that such solutions can be given safely in large volumes with only a minimum requirement for monitoring. The lactate in Hartmann's or lactated Ringer's solution may help to control acidosis as it is metabolized to bicarbonate by the liver.
- Hypertonic saline (7.2% NaCl) has found favour recently in the emergency treatment of dehydrated horses. The precise physiological mode of action is uncertain but the net result of intravenous administration of such fluid is redistribution of fluid from the extra-cellular space into the circulation. Small volumes of hypertonic saline result in a rapid improvement in cardiovascular function. In a clinical situation the infusion of 2.5 litres of 7.2% NaCl would be expected to improve a horse's PCV by approximately 10 percentage points in 30 minutes. The major disadvantage of this type of emergency fluid therapy is that it **must** be followed by large volumes of isotonic fluids within two hours. For this reason, careful monitoring of hydration status and the availability of isotonic fluids are essential prerequisites to such therapy.

Hypertonic saline would be the fluid of choice in this case as long as the appropriate monitoring facilities and follow-up fluids were available. The rapid improvement in cardiovascular status leading to improvement in cardiac output and peripheral perfusion should win the approval of your anaesthetist colleague. Administration of a colloid preparation (eg. plasma) would enhance the effect of crystalloids

95 i. Inguinal and scrotal hernia.
ii. Most inguinal and scrotal hernias will resolve spontaneously and surgery is rarely necessary. Conservative therapy is usually sufficient, and involves palpation and manual reduction of the hernia twice a day. In the rare event of herniated intestine becoming incarcerated and strangulated, the hernia becomes turgid and irreducible, and the foal will usually show signs of abdominal pain. In this event, surgical interference is required (inguinal herniorrhaphy, usually combined with bilateral castration).

96 i. Non-steroidal anti-inflammatory toxicity. CBC to investigate evidence of inflammation secondary to gastrointestinal mucosal damage, serum total protein and albumin to evaluate gastrointestinal protein loss, abdominocentesis to evaluate the severity of intra-abdominal inflammation.
ii. Right dorsal colon.

97–99: Questions

97 A four-day-old Thoroughbred filly is presented weak and lethargic with moderate muscle fasciculations. The heart rate is 120bpm, respiratory rate 28bpm and the rectal temperature is 37.3°C. The abdomen is distended and appears pendulous. The filly has been noted to urinate a normal stream since birth. An abdominal ultrasound examination is shown in **97**.

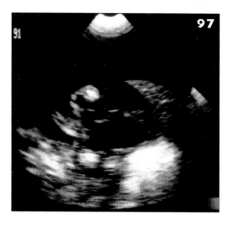

i. What is the likely diagnosis?
ii. What are the sources of the fluid accumulating in the abdomen?
iii. What would be the expected laboratory findings?
iv. How can the diagnosis be confirmed?
v. What is the recommended treatment?

98 A mare was examined for mild colic of two days duration. On clinical examination there were no remarkable findings except severe jaundice of the mucous membranes. A CBC was normal but blood chemistries revealed a marked elevation in GGT, high normal AST and increases in both conjugated (54.7µmol/l; 3.2mg/dl) and unconjugated bilirubin (92.3µmol/l; 5.4mg/dl). A liver biopsy showed bile duct proliferation with mild to moderate perilobular fibrosis. The mare was grazing pasture composed almost entirely of the plant shown in **98**.

i. What is the most likely diagnosis in this case?
ii. What is the prognosis?

99 To confirm the diagnosis of glomerulonephritis, immunofluorescent examination of a section of renal cortex would be desirable.
i. Assuming ultrasound examination is available, which kidney would it be preferable to biopsy in a pony with bilateral renal disease (**49**), and why?
ii. What instrument would be used to collect the biopsy?
iii. What should the renal tissue be placed in for routine histopathology and immunohistochemical studies?

97–99: Answers

97 i. Uroperitoneum.
ii. Urine leakage from the ureters, the bladder (neck or dorsal regions), or the urachus.
iii. Moderate elevations in blood urea nitrogen and creatinine, low sodium, high potassium and variable bicarbonate levels.
iv. A combination of the laboratory findings and the ultrasound evaluation is usually sufficient for the diagnosis. A peritoneal fluid creatinine at least twice the level found in the bloodstream would confirm the presence of uroperitoneum.
v. Surgical repair. Excessive potassium levels represent a greater anaesthetic risk of cardiac arrest. Pre-surgical bicarbonate, fluid therapy and drainage of the peritoneal fluid may be necessary in severe cases. Oral sodium polystyrene can also be given pre-operatively.

98 i. The most likely diagnosis is alsike clover poisoning. To help confirm the diagnosis, blood was drawn from other horses on the pasture and all had elevations in GGT, although they were asymptomatic. All the horses had normal GGT levels within three weeks of being removed from the pasture.
ii. The prognosis was good in the clinically affected mare since she was not exhibiting signs of hepatoencephalopathy, the hepatocellular enzyme AST was within normal range, the liver biopsy did not reveal severe fibrosis and the mare could be removed from the source of the toxin.

99 i. The right kidney is biopsied preferably in horses with bilateral renal disease. The right kidney is closer to the body wall than the left, is less moveable than the left kidney, and is more easily visualized with ultrasound; also, there are no vital structures between the body wall and right kidney (in contrast to the left which has the spleen positioned between the kidney and the body wall).
ii. The preferred instrument is an 11.25–15cm, 14–16G biopsy needle, such as a Tru-Cut biopsy needle from Baxter Healthcare Corporation, Valencia, Calif., USA.
iii. For routine histopathology the kidney should be placed in 10% formalin. For immunohistochemical staining the kidney should be snap frozen at −70°C or placed in Michel's transport solution.

100 & 101: Questions

100 A mare is referred in April with a history of persistent oestrus. The owner complains that the mare has been showing mild signs of oestrus, ie. everting the clitoris when in the company of strange geldings, since she was purchased in December. Since March she has been showing strong signs of oestrus, ie. seeking the company

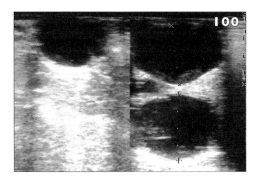

of other horses, been difficult to handle, backing up to other horses in the field and squirting urine. The owner wants to compete the mare over the summer and requests that the mare is ovariectomized. The mare is teased with a stallion and does show strong signs of oestrus. An ultrasound scan of the ovaries is shown (**100**). The uterus shows marked oedema. How are you going to manage this case?

101 A two-day-old Thoroughbred colt was observed to posture frequently in the manner depicted in **101a**. A sonogram of the caudoventral abdomen, longitudinal view, was obtained (**101b**).
i. What is this foal exhibiting, and what are the differential diagnoses?
ii. What is the diagnosis based on the sonographic exam?

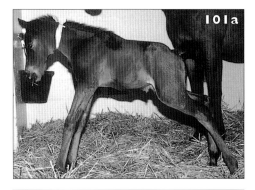

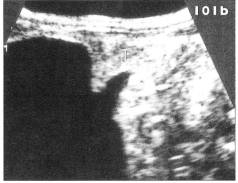

81

100 & 101: Answers

100 It is extremely important when this case is worked up that the behaviour is checked with a stallion (preferably) or strange gelding. Several types of behaviour are often misinterpreted as oestrus behaviour:
- Tail lifting and frequent urination can occur in mares with back or hindlimb pain, or urogenital infection, obstruction or incontinence.
- Submissive mares when placed in stressful conditions can squat, squirt urine, swish the tail and wink, especially when in close contact with humans or other horses. Mares exhibiting guarding behaviour can show similar signs. This type of behaviour can be associated with racing mares in starting stalls or when being saddled, and leads to such mares being termed 'nymphomaniac'. Again, testing with a stallion will distinguish these signs from true oestrus.
- Stallion-like behaviour in mares with a granulosa cell tumour can be misinterpreted by owners as excessive oestrus. However, granulosa cell tumours can also be associated with continuous or intermittent oestrus behaviour.

Ovarian palpation and scanning need to be performed in conjunction with behavioural observations. The ultrasound scan of the ovaries of the mare in the present case demonstrates three large follicles. It is likely that this mare is in spring transition, ie. in the period between anoestrus and normal cyclicity. During this time it is common for follicles to reach >30mm and then regress. In association with this increased follicular activity, mares can show prolonged periods of oestrus. The usual treatment for a mare with this history is to administer a course of oral progestagens (altrenogest) for 10–15 days. This will suppress oestrus behaviour and may hasten the onset of normal cyclicity when it is withdrawn. If the mare's performance is still poor in association with the periods when she is in oestrus, she can remain on progestagen treatment over the summer show period. This therapy will suppress oestrus behaviour, but not necessarily ovulation. There is very little detailed information which compares behaviour before and after bilateral ovariectomy. Because of the sparse objective data available, it seems preferable in all cases to try to avoid ovariectomy if possible, by assessing response to hormonal therapy first.

101 i. Stranguria. Differentials include ruptured bladder or urachus (with uroperitoneum), bladder atony and urachal diverticulum.
ii. Grossly intact bladder with a urachal diverticulum. There is no free abdominal fluid as would be seen with uroperitoneum (see **97**).

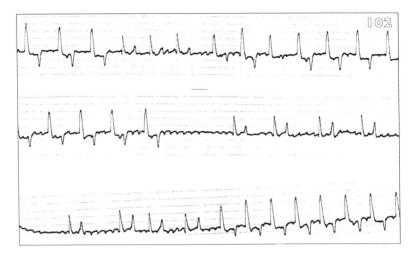

102 This ECG (102) was obtained from a horse with acute onset of exercise intolerance. Auscultation revealed an irregular rhythm with a rate of 40–50bpm, but periods of a rapid (about 80bpm) regular rhythm were also heard. There were no murmurs.
i. What is the rhythm?
ii. What other diagnostic procedure(s) should be performed?
iii. What would be a logical treatment regimen for this horse assuming the test(s) in the answer to ii is/are normal?

103 You are asked to examine a seven-year-old Warmblood mare with regard to progressive bilateral swellings of the distal forelimbs which have been present for several months. Both forelimbs have firm, non-painful enlargements of the distal 3rd metacarpal bones and the proximal and middle phalanges. The rectal temperature is 38.0°C. A haematological evaluation revealed a leucocyte count of $6.9 \times 10^9/l$ (n = $7.0–10.0 \times 10^9/l$) with 79% neutrophils and 21% lymphocytes. Serum levels of alkaline phosphatase are mildly elevated (351u/l; n = 250–300u/l).
i. What is the likely diagnosis?
ii. What is the aetiology of this condition?
iii. How could you confirm the diagnosis?
iv. How would you treat this horse?
v. Can regression of this condition be expected?

102 & 103: Answers

102 i. Atrial fibrillation with paroxysmal ventricular tachycardia. Notice that there are no P waves, and that there are QRS-T wave forms of two types. The large QRSs with negative T waves are ventricular in origin.
ii. An echocardiogram should be done to determine if any structural heart disease, such as a cardiomyopathy/myocarditis, exists. A CBC and chemistry panel were done in this horse to determine if any systemic disorders or electrolyte disturbances were present. Serum cardiac isoenzymes were also evaluated. All these tests were normal.
iii. The ventricular tachycardia was not considered to be life threatening because of the rate (less than 100bpm), that it was uniform (one foci in ventricle), and because the horse was asymptomatic at rest. Atrial fibrillation may resolve spontaneously within 24 hours, therefore re-evaluation after 24 hours of onset of signs would be appropriate. The drug of choice for treatment of this arrhythmia is quinidine. It is effective in treating both ventricular arrhythmias and atrial fibrillation. After 24 hours, the ventricular tachycardia had resolved, but the atrial fibrillation was still present. The atrial fibrillation was successfully converted to sinus rhythm after several doses of quinidine sulphate had been given by nasogastric tube. Quinidine gluconate administered in intravenous boluses may also be used to treat atrial fibrillation of under two weeks duration in the horse.

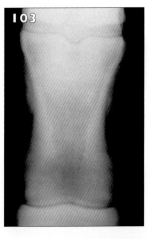

103 i. The likely diagnosis is hypertrophic osteopathy (Marie's disease). This disease is characterized by the symmetrical proliferation of vascular connective tissue and subperiosteal bone at various sites, in particular the distal long bones. The forelimbs, hindlimbs or both can be involved. Hypertrophic osteopathy is a rare disease in the horse.
ii. The precise cause of hypertrophic osteopathy is unknown. It is usually found in association with intrathoracic diseases, especially inflammatory lesions or neoplasia. However, it can also occur secondary to diseases in other locations. Reports of hypertrophic osteopathy in horses have included cases associated with tuberculosis, pulmonary abscesses, pulmonary neoplasms, ovarian neoplasms with or without pulmonary metastasis, pituitary pars intermedia adenoma, rib fracture and pulmonary infarction.
iii. A thorough clinical examination including rectal palpation is required. Radiography of affected limb bones reveals bilateral periosteal new bone formation (103). In addition, radiographic examination of the thorax should be performed in an attempt to identify any intrathoracic disease. An elevation of serum alkaline phosphatase levels is variably found, indicating increased osteoblastic activity associated with periosteal proliferation.
iv. Identify and treat the underlying disease process.
v. If the underlying disease process is eliminated, almost total regression of the hypertrophic osteopathy can be expected.

104–106: Questions

104 You are performing a routine reproductive examination on a mare. On removal of your gloved hand from the rectum you notice some fresh blood on the glove (**104**). Describe how you would deal with this situation from both a clinical and a legal perspective.

105 You are asked to examine a seven-year-old crossbred gelding with progressive weight loss (approximately 100kg) and intermittent pyrexia (rectal temperature up to 38.6°C) of two months duration. On examination the gelding is thin, and shows muscle wasting and ventral abdominal oedema (**105**). Faecal consistency is normal. Clinical pathological examinations reveal a

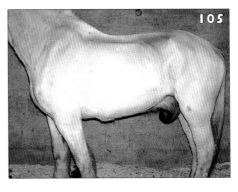

mild normocytic, normochromic anaemia, hyperfibrinogenaemia (plasma fibrinogen 6.2g/l), hypoalbuminaemia (serum albumin 15g/l) and hypoglobulinaemia (serum globulin 22g/l). The white blood cell count and serum biochemistry results are within normal limits. Peritoneal fluid analysis and urinalysis are normal. A Coggins test is negative. Thoracic and abdominal ultrasonography are unremarkable.
i. What type of disease or syndrome do you suspect?
ii. How could you confirm your suspicion?
iii. What further diagnostic procedures are indicated?
iv. What are the commonest pathological forms of disease associated with this syndrome?

106 What abnormalities of blood gas analysis would you expect to find in the following states:
i. Metabolic acidosis with acidaemia?
ii. Metabolic alkalosis with alkalaemia?
iii. Respiratory acidosis with acidaemia?

104–106: Answers

104 The (thankfully rare) occurrence of rectal tears is both professionally embarrassing and potentially expensive if the client successfully sues the veterinary surgeon involved. The successful clinical outcome of these iatrogenic injuries and the minimization of legal liability both involve:
- Admission that a problem has occurred.
- Communication with the client.
- Active first aid measures and management.

The first step in managing this situation is assessment of the nature of the injury. The patient should be sedated to facilitate further examination, and in many cases epidural anaesthesia is indicated to prevent straining and enlargement of the tear. The rectum should be evacuated manually and a careful palpation and visual inspection via a speculum be performed, paying particular attention to the dorsal wall 20–30cm from the anus. Rectal tears are classified according to their depth and degree of tissue involvement. Prognosis is directly related to this classification. Minor abrasions of the rectal wall involving just the mucosa and submucosa are classified as Grade I tears. Management of these cases is usually successful if further examinations *per rectum* are prohibited and the animal monitored for signs of discomfort and colic over the ensuing 12–24 hours. This may be most easily achieved at a hospital facility. Monitoring of peritoneal fluid for elevations in neutrophils and protein content may be helpful diagnostically and prognostically. Grade II tears involve just muscularis and are rarely recognized clinically. Tears that extend through mucosa, submucosa and muscularis are classified as grade III. The priority with this type of injury is to minimize faecal entry into the tissue layers; this will cause enlargement of the defect and possibly result in a full thickness tear (grade IV lesion) with passage of faecal material directly into the peritoneal cavity. This aim is most easily accomplished by packing the rectum (after epidural anaesthesia) with cotton sheeting soaked in povidone iodine solution. Antibiotic cover should be commenced at once. Having taken these remedial steps, the horse can be safely transferred to a hospital facility for further assessment, colostomy or other surgical procedure. The prognosis for grade III and IV tears is guarded to poor, but rapid, appropriate first aid measures have been shown to improve significantly the success rate of surgically managed cases.

105 i. The clinical and laboratory findings are suggestive of small intestinal malabsorption.
ii. Small intestinal malabsorption is most easily confirmed by an oral sugar absorption test, such as the oral glucose or xylose absorption tests.
iii. Small intestinal wall and mesenteric lymph node biopsies.
iv. Alimentary lymphosarcoma, granulomatous enteritis, eosinophilic gastroenteritis, eosinophilic granulomatous enteritis, lymphocytic plasmacytic enteritis.

106 i. Low pH, decreased plasma bicarbonate, normal or low $PaCO_2$.
ii. High pH, increased plasma bicarbonate, normal or increased $PaCO_2$.
iii. Low pH, increased $PaCO_2$, normal or increased plasma bicarbonate.

107–109: Questions

107 Shown are an ECG recorded at rest (**107a**) and one recorded during trotting exercise (**107b**) from the same horse.

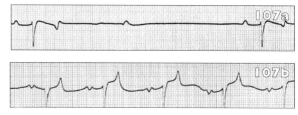

i. What do the ECGs demonstrate?
ii. What is the mechanism which causes this arrhythmia?
iii. Which techniques are used to record the ECG during exercise?

108 You are asked to examine a four-month-old colt for what the owner describes as a lump on its belly. The foal has no history of previous illness. Your physical examination reveals a mass around the umbilicus (**108**). Upon palpation, the swelling is soft and the contents can be easily reduced into the abdomen.

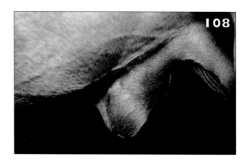

i. What is your diagnosis?
ii. What ancillary diagnostic tests might be considered in addition to physical examination?
iii. What treatment plan would you recommend?

109 An adult Thoroughbred horse has had a history of weight loss and mild colic for the previous eight weeks. Euthanasia was performed. The stomach has been opened and is shown here (**109**) with a very roughened, coarsely granular and thickened squamous portion of mucosa. Scattered *Gastrophilus intestinalis* larvae are present on this surface. The glandular surface is relatively unaffected. What is the most likely diagnosis?

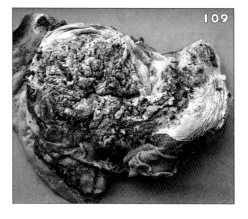

107–109: Answers

107 i. **107a** shows second degree atrioventricular block, characterized by a P wave which is not followed by a QRS complex. **107b** shows that the arrhythmia has disappeared at higher heart rates.
ii. Second degree atrioventricular block is a common finding in normal horses. It is vagally-mediated and is a means by which the horse can regulate its cardiac output and blood pressure. The vagus innervates both the sinoatrial and the atrioventricular nodes and, as vagal tone increases, this slows and may intermittently prevent conduction through the atrioventricular node, causing second degree atrioventricular block. When sympathetic tone increases, for example during exercise or excitement, the arrhythmia is abolished, as is illustrated in **107b**. If second degree atrioventricular block persists during higher heart rates or third degree (complete) atrioventricular block is present, this is abnormal; it is usually due to inflammation or fibrosis of the atrioventricular node or severe electrolyte disturbances, notably hyperkalaemia secondary to renal failure or uroperitoneum.
iii. Exercising ECGs can be recorded by attaching conventional ECG leads while exercising the horse on a treadmill or using a radiotelemetry system (a transmitter is attached to the horse and sends a signal to a monitor at a distant point by radio).

108 i. An umbilical hernia, considered to be the most common form of abdominal hernia in horses. These defects are considered developmental. Omphalophlebitis and increased intra-abdominal pressure have been proposed as factors that contribute to failure of the umbilicus to close.
ii. Diagnosis of umbilical herniation is based on palpation, auscultation of the region and ultrasonography. Palpation of an uncomplicated umbilical hernia will not elicit pain and the hernia contents will be easily reduced within the abdomen. If bowel is herniated, auscultation may reveal intestinal sounds. Ultrasonography of the umbilical swelling can identify herniated bowel and abnormalities of internal umbilical structures.
iii. Small umbilical hernias (<5cm) usually close as the foal ages and do not generally cause intestinal obstruction and colic. Owners should be instructed to reduce the hernia daily and watch for changes in the size of the hernia sac and elicitation of pain during palpation. If small hernias do not close by one year of age, herniorrhaphy is recommended. Larger defects (>10cm) will not generally close spontaneously and should be closed by herniorrhaphy.

109 Squamous cell carcinoma of the stomach.

110–112: Questions

110 An eight-year-old Warmblood is examined because of lip droop (**110**), muzzle deviation to the right, right sided ear droop, decreased palpebral size on the right, decreased tail tone, faecal impaction, absence of anal reflex, dilated anus, and perineal and penile analgesia. The clinical signs were first noted two weeks previously and have been progressive. Vital signs were normal except for a rectal temperature of 38.3°C.
i. What is the diagnosis?
ii. What diagnostic procedures could be performed to help establish the diagnosis?
iii. Which cranial nerve is involved and what is the most serious complication associated with paralysis of this nerve?

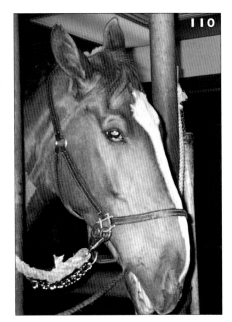

111 A six-year-old Quarterhorse stallion is examined because of rapidly progressing facial swelling, particularly of both masseter muscles, to the extent that the conjuctiva of both eyes were protruding from the orbit. A urine sample was dark and strongly positive for occult blood (**111**).
i. What is the most likely diagnosis in this case?
ii. What is the treatment?

112 i. How do you diagnose hypothyroidism?
ii. How would you treat hypothyroidism?

110–112: Answers

110 i. Neuritis of the cauda equina (polyneuritis equi).
ii. A lumbosacral spinal tap could be performed but is often difficult in these cases because of scarring of the dura around the cord. If fluid is obtained, it typically has a high protein content and pleocytosis with a predominance of lymphocytes and macrophages. Antibodies may be detected in the CSF and serum for myelin protein antigens but this may not be specific for neuritis of the cauda equina. In fact the clinical signs, signalment and history are all that are required to make the diagnosis.
iii. The facial nerve is affected in this case. All branches of the nerve are involved, there is no vestibular dysfunction and other neurological signs are present, so it can be assumed that the facial nerve nuclei are affected. The most serious complication from facial paresis/paralysis in the horse is exposure keratitis and corneal ulceration.

111 i. The diagnosis is masseter myopathy caused by a severe deficiency of selenium.
ii. Treatment is intramuscularly administered selenium, intravenous polyionic fluids and non-steroidal anti-inflammatory drugs.

112 i. Hypothyroidism can be diagnosed in horses on the basis of measuring the T_4 concentration in plasma and the response to the administration of thyroid stimulating hormone (TSH) or thyroid releasing hormone (TRH). After taking a basal morning blood sample, 5 or 2.5iu of (bovine) TSH is administered intravenously. Blood should be collected four hours later. In normal horses, T_4 will be increased to more than twice the basal value. The same results can be obtained in normal horses when TSH is replaced by 0.5–1.0mg of TRH i/v.
ii. Synthetic thyroxine at a dose of 20μg/kg bodyweight orally maintains normal thyroid hormone concentrations for 24 hours. In cases of euthyroid sick syndrome, a thorough clinical examination including rectal palpation is indicated.

113 & 114: Questions

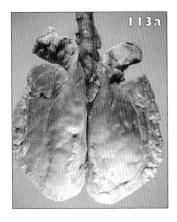
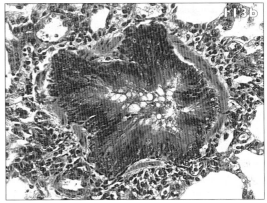

113 Shown are the gross (**113a**) and histological (**113b**) post-mortem appearances of the lungs of a horse affected by long-standing chronic obstructive pulmonary disease (COPD).
i. What abnormalities are present?
ii. What is the pathophysiology of COPD?
iii. Some horses may demonstrate clinical signs similar to COPD, but occurring in the summer time when the horses are pastured and without any access to hay or straw. What is this disease and how can it be treated?

114 This four-year-old horse (**114**) suddenly developed widespread, non-pruritic skin lesions over its entire trunk. The hair was clipped to examine the lesions closely. After examination, the horse was given multiple intravenous injections of dexamethasone.
i. List the two most likely disorders to cause the lesions seen.
ii. If the lesions disappeared with treatment but recurred when the treatment was discontinued, what is the most likely diagnosis?
iii. If the lesions did not disappear with treatment, what is the most likely diagnosis?

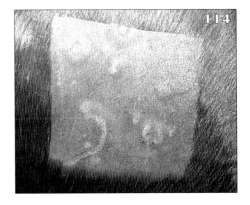

113 & 114: Answers

113 i. There is overinflation of the lungs with indentations of the surface caused by pressure from the ribs. Emphysematous change is present in the apical lobes and along the edges of the diaphragmatic lobes. In many cases of COPD the lungs will appear grossly normal, and emphysema only develops following long-term uncontrolled disease.

The histological section shows bronchiolar epithelial cell hyperplasia, peribronchial infiltration by neutrophils, lymphocytes and plasma cells, and inflammatory cell accumulation in the airway lumen. Chronic bronchiolitis, sometimes with alveolar overinflation, is the most consistent histological abnormality in COPD.

ii. Horses affected by COPD generally have an increased airway resistance and decreased dynamic compliance, and a prolongation of nitrogen washout. All of these changes are compatible with diffuse lower airway obstruction. The exacerbated regional differences in small airway resistance and lung compliance result in an uneven distribution of ventilation. The resulting inequalities in ventilation/perfusion ratio may in turn result in hypoxaemia. The increase in total pulmonary resistance and the impaired gas exchange decrease the exercise tolerance of affected horses. Diffuse lower airway obstruction of COPD is in part caused by plugging of airways with mucus and exudate, and hyperplasia of the airway epithelium. In addition, airway smooth muscle contraction plays a major role.

iii. Summer pasture-associated obstructive pulmonary disease (SPAOPD). This is similar clinically to COPD but differs in that it occurs in grazing horses in the summer. The condition is well known in warm humid areas of south-eastern USA. A similar syndrome is being recognized increasingly in the UK. SPAOPD affects mature horses of all breeds. The condition resolves spontaneously in the winter (unless the horse is concurrently affected by COPD) and recurs on an annual basis in the summer. In the USA, clinical signs of SPAOPD are most commonly seen in the summer and early autumn, whereas in the UK disease appears most commonly in the spring and early summer. The clinical features of SPAOPD are similar to COPD and include tachypnoea, expiratory dyspnoea, coughing and nasal discharge. In some cases the dyspnoea is severe, and the affected horse may be unable to work. Auscultation of the chest reveals inspiratory and expiratory crackles and wheezes. Mucus sounds are commonly audible over the trachea. Tracheal aspirates and bronchoalveolar lavage samples show neutrophilia and increased amounts of mucus.

Treatment includes removal of the horse from the offending environment. In many cases this will involve stabling the horse and feeding hay. If the horse is concurrently affected by COPD, stable management should be adjusted to eliminate hay and straw from the environment. Drug therapy is frequently necessary in horses showing severe clinical signs. Bronchodilators and corticosteroids (alone or in combination) are helpful, but the potential side-effects of corticosteroid therapy must be considered.

114 i. Urticaria, erythema multiforme.
ii. Urticaria.
iii. Erythema multiforme (very rare).

115–117: Questions

115 You are asked to examine a three-year-old Thoroughbred mare with a history of intermittent coughing. The horse has a normal rectal temperature. No abnormalities are found on bilateral auscultation and percussion of the thorax. The animal coughs on superficial palpation of the larynx. There is no history of poor performance. Endoscopic examination of the nasopharynx reveals numerous nodules scattered over the walls and roof (**115**).

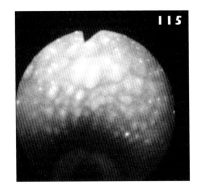

i. What is your diagnosis?
ii. Describe the usual course of the disease.
iii. What is the aetiology of this disease?
iv. How would you treat this horse?

116 Shown (**116a, 116b**) are the clinical findings of septicaemia in the foal associated with vascular compromise. Give at least one physical finding or clinical sign of sepsis for each of the following body systems:
i. Ocular.
ii. Integumentary.
iii. Locomotor.
iv. Cardiac.
v. Respiratory.
vi. Gastrointestinal.
vii. Urogenital.
viii. Neurological.

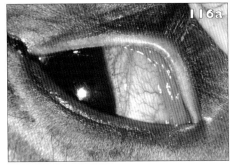

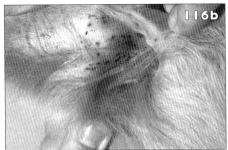

117 List plants other than alsike clover (see **98**) that are known to produce liver disease and failure in the horse.

115–117: Answers

115 i. Pharyngeal lymphoid hyperplasia (pharyngitis).
ii. Many authors consider pharyngeal lymphoid hyperplasia to be a normal physiological finding. There is a relationship between the prevalence of pharyngeal lymphoid hyperplasia and age. All animals older than three months develop a certain degree of pharyngeal lymphoid hyperplasia. At least 70% of all two-year-old horses have a degree of this condition. In older animals the prevalence gradually decreases, and in animals older than seven years very few lymphoid follicles are seen. The condition may cause intermittent coughing, and coughing is easily elicited by laryngeal palpation. Sometimes, especially in horses with extensive lymphoid hyperplasia, a respiratory sound may be heard during exercise. Nevertheless, pharyngeal lymphoid hyperplasia is not considered to be a cause of poor performance.
iii. The precise aetiology of pharyngeal lymphoid hyperplasia is uncertain, and is probably multifactorial. The lymphoid hyperplasia probably represents an immunological response in the pharyngeal lymphoid tissue to novel bacterial, viral and other environmental antigens.
iv. In most cases the best therapy is to do nothing. Most horses are not seriously affected by the condition, and as the lymphoid hyperplasia tends to regress with age, treatment is not often necessary. If the horse appears to be clinically affected by the condition, there are several treatment options:
- A period of rest.
- Stable management changes to reduce exposure to stable dust and ammonia.
- (Local) anti-inflammatory treatment.
- Cryosurgery.
- Electro- or chemo-cauterization.

Cryosurgery and cauterization may be questionable techniques. There are several structures in close proximity to the dorsal wall of the pharynx, eg. nerves, that may be damaged. Formation and retraction of scar tissue may be more deleterious than the original condition. Although several species of bacteria can often be isolated from cultures of the pharynx of affected horses, their significance is questionable. Therefore, antibacterial therapy is not indicated.

116 i. Hypopyon, iritis.
ii. Anasarca.
iii. Joint effusion and/or septic arthritis.
iv. Tachycardia.
v. Polypnoea.
vi. Ileus, colic, diarrhoea.
vii. Anuria.
viii. Seizures, depression.

117 There are several plants that contain pyrrolizidine alkaloid which causes hepatic fibrosis. These include: *Senecio* spp. (ragwort, groundsel, tar weed), *Amsinckia* (fiddleneck), *Crotolaria* (rattlebox), *Heliotropium* (common heliotrope) and *Cynoglossum* (hounds tongue). Some plants that contain saponins (eg. *Panicum* spp. [kleingrass, fall panicum]) may cause biliary hyperplasia, hepatic fibrosis and occasionally liver failure.

118–120: Questions

118 A four-year-old, Thoroughbred-cross gelding is under investigation for recurrent episodes of colic and gradual weight loss. The problem began six weeks previously and has resulted in eight episodes of colic, each one lasting less than an hour and responding to medical treatment with spasmolytics. Rectal examination is uninformative, either during the colic episodes or between them. Blood has recently been submitted for haematology and biochemistry and the following results received (normal ranges in brackets): Haematology: RBC, $9.7 \times 10^{12}/l$ ($8.5–12.5 \times 10^{12}/l$); WBC, $18.7 \times 10^9/l$ ($6.0–12.0 \times 10^9/l$); neutrophils, $12.2 \times 10^9/l$ ($2.7–6.7 \times 10^9/l$); lymphocytes, $5.8 \times 10^9/l$ ($1.5–5.5 \times 10^9/l$); monocytes, $0.5 \times 10^9/l$ ($0–0.2 \times 10^9/l$); eosinophils, $0.2 \times 10^9/l$ ($0.1–0.6 \times 10^9/l$).
Biochemistry: creatine kinase, 148u/l (<50u/l); AST, 190u/l (80–250u/l); GGT, 14u/l (<40u/l); total protein, 82.5g/l (62.5–70g/l); albumin, 28.5g/l (27–36.5g/l); globulin, 54g/l (17–40g/l); fibrinogen, 5.5g/l (0–4g/l).
How would you interpret these results, what would be your list of differential diagnoses and how would you proceed with the investigation of this case?

119 During a pre-purchase (soundness) examination of a six-year-old Quarterhorse, the left eye appears normal. However, the right eye appears as in **119**.
i. What is this lesion?
ii. The horse is performing adequately. Do you recommend purchase?

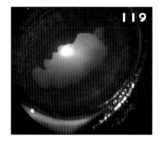

120 A 27-year-old mixed breed mare is examined because of progressive weight loss, haematuria throughout urination and, frequently, large clots at the end of urination. On physical examination the mucous membranes are abnormally pale (**120**) and the heart rate is 64bpm. Rectal examination is normal. CBC and serum chemistries reveal the following abnormalities: PCV, 14%; total protein, 50g/l; calcium, 3.52mmol/l. Urinalysis is normal except for red discolouration and a large number of red blood cells. Endoscopy of the bladder is normal except for red urine coming from the left ureter.

i. What would be your next diagnostic procedures?
ii. What is the most likely diagnosis?

118–120: Answers

118 The haemogram of this horse shows a marked leucocytosis which is due to a neutrophilia. This is evidence of an active, acute inflammatory response. The biochemical profile further supports this interpretation as fibrinogen (an acute phase protein) is elevated, as is globulin. This may be associated with the recent history of colic and weight loss if the inflammatory focus is within the abdomen. One way of localizing the inflammatory response is to take a peritoneal fluid sample for cytology, and this would seem to be the next logical step in this investigation. The peritoneal fluid from this particular horse was turbid, with a nucleated cell count of 70×10^9/l (reference range $0–5 \times 10^9$/l). On the basis of these findings it was decided that an exploratory laparotomy should be performed. Bearing in mind the suspected inflammatory basis of this problem and its localization to the peritoneal cavity, a list of differential diagnoses should include:
- Intra-abdominal abscessation as was found in the case described at laparotomy and subsequent post-mortem examination.
- Localized peritonitis (eg. foreign body penetration of gut).
- Adhesions involving small intestine (due to a variety of possible causes, including intra-abdominal abscessation and localized peritonitis).
- Chronic intussusception.
- Necrotic intra-abdominal neoplasm.
- Enterolith abrasion/perforation of large colon.
- Sand accumulation causing large colon inflammation.

119 i. Cataract due to prior uveitis and subsequent posterior synechiae.
ii. No. Subsequent episodes of uveitis would be likely in the left eye, and very possibly as a future problem in the right eye. Over 50% of horses will have uveitis as a bilateral condition in time. Traumatic uveitis may have caused the left eye lesion, but unless you actually attended the trauma, medicolegal logic dictates a failed soundness examination.

120 i. An ultrasound examination of the kidneys should be performed as should an abdominocentesis. These are performed to detect abnormal renal masses (eg. tumour) or neoplastic cells in the abdominal fluid.
ii. The most likely diagnosis is renal carcinoma.

121 & 122: Questions

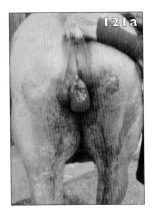
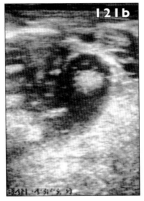
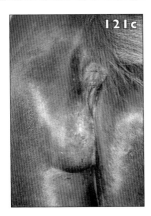

121 A post-partum mare is presented with a history of delivering a large foal with some difficulty 12 hours previously. The mare passed her placenta within 30 minutes of foaling but has appeared to be colicky since then and a bloody structure is present at her vulval lips (**121a**). Observation of the mare shows her to be apprehensive, looking at her flanks and shifting weight from one hindlinb to the other. Clinical examination reveals her heart rate to be 70bpm. Mucous membranes are slightly pale. It is noticeably difficult to outline the edges of the uterus by palpation per rectum and there appears to be a mass over the left horn and left side of the vagina. An ultrasound scan is performed (**121b**) and this structure is outlined. A week later a bulge appears lateral to the left vulval lip (**121c**).
i. What is the structure visible at the vulval lips most likely to be?
ii. How are you going to treat it?
iii. What is likely to account for the elevated heart rate, mild colic, slightly pale membranes and the scan image? If this presents in a more severe form, how should the case be managed?

122 A two-year-old Standardbred is examined because of lethargy, weight loss, diminished appetite and bilateral firm swelling of the upper eyelids. Clinical examination is unremarkable except for the thickened, non-painful swellings of the eyelids.
i. What is the most likely diagnosis?
ii. How can it be confirmed?
iii. Which results from chemistry and immunoglobulin panels might be helpful in the diagnosis?

121 & 122: Answers

121 i. A vaginal haematoma.
ii. Small haematomas which do not protrude through the vulval lips do not need treatment. A Caslick's vulvoplasty can be performed to prevent pneumovagina if the mare is straining. Local application of ice-packs or hydrotherapy can help to limit the size of large haematomas during the first few hours after foaling. These haematomas can be drained 2–3 days post-partum after the clot has organized. Systemic antibiotic cover should be administered to the mare.
iii. Rupture of the uterine or utero-ovarian artery can present with these signs when the haemorrhage is contained within the two layers of the broad ligament. Artery rupture tends to occur in older mares possibly due to age-related degenerative changes in the blood vessels. Increased parity may also predispose to rupture. Most will occur at parturition – rarely before. It is estimated that approximately 50% of mares show no clinical signs in association with uterine artery rupture. In most of these mares the haematoma will organize and shrink over the next few months. However, if the bleeding starts again, the broad ligament may rupture and the mare will bleed into the abdomen which usually results in death.

Alternatively, mares with artery rupture may fail to show signs until they develop weakness or anaemia. The colicky signs associated with haematoma formation can be mistakenly diagnosed as normal placental expulsion and uterine involution. It is advisable to keep the mare quiet in a darkened loose box under heavy sedation. Exciting the mare can worsen the bleeding and cause broad ligament rupture. Analgesics may be administered to treat the pain associated with broad ligament distension. Oxytocin should not be given because the associated uterine contractions can rupture the broad ligament haematoma. Prophylactic antibiotic therapy should be administered for up to a week after the haemorrhage. If the haemorrhage is not contained within the broad ligament, the mare will suffer from haemorrhagic shock. Treatment is frequently unsuccessful. Severe haemorrhage may cause death without a decline in PCV.

If the PCV from more prolonged bleeding falls below 20%, whole blood transfusion is indicated.

In the case described here, bleeding was presumably from a caudal artery such as the uterine branch of the vaginal artery. The haemorrhage presumably tracked back along the wall of the vagina, where it was detected by ultrasonography five days after foaling, and then appeared as a swelling in the left perineal region.

122 i. Lymphosarcoma.
ii. A fine needle aspirate of the swollen lid should confirm the diagnosis.
iii. An elevated serum calcium would be supportive as would an abnormally low serum IgM. The sensitivity and specificity of both tests for diagnosing lymphosarcoma are not high.

123 & 124: Questions

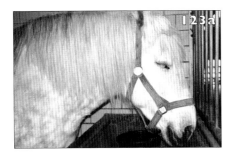
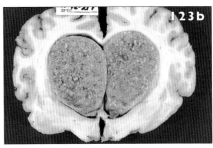

123 A 10-year-old Percheron mare (**123a**) is examined because of profound depression. The mare is kept in a pasture with one other mare and not closely observed, making time of onset of clinical signs unknown. There was no evidence of trauma, no fever, no detectable cervical pain, no muscle atrophy and no cranial nerve deficits. The mare preferred to hold her head down. She was reluctant to move and was especially reluctant to turn to the right. When forced to walk she had a stiff gait but was not ataxic. Spinal fluid collected from the lumbosacral space had elevated protein (147g/l), increased RBCs (1679 cells/µl) and a pleocytosis (27 nucleated cells/µl), of which macrophages were the predominant cell type. Euthanasia was carried out in view of the poor prognosis. On transverse sectioning of the brain, two large firm masses were found within and expanding both ventricles (**123b**).
i. What is the diagnosis?
ii. Why is there an increased number of macrophages in the CSF?
iii. What are other differentials that should have been considered in this case?

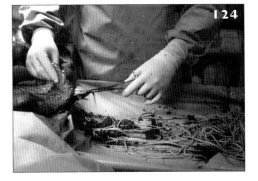

124 Exploratory laparotomy was carried out on a 14-month-old Standardbred colt on the evidence of physical findings consistent with small intestinal obstruction. At surgery the findings were those illustrated in **124**.
i. What is the diagnosis?
ii. What advice would you give to the owner of the colt?

123 & 124: Answers

123 i. Cholesterol granuloma (cholesterolaemic granuloma; cholesteatoma). These are most often incidental findings at necropsy but when large enough to compress neuronal tissue or obstruct CSF drainage, they may cause neurological signs.
ii. Haemorrhage into the CSF, which occurred in this case because of the large size of the mass, often stimulates macrophage influx in the CSF in order to clean up the red cells.
iii. Abscess, neoplasia, mouldy corn poisoning, viral and parasitic encephalomyelitis, equine infectious anaemia and trauma.

124 i. Small intestinal impaction with *Parascaris equorum*. Ascarid infection is common in equine animals less than two years old, after which a strong immunity to this parasite develops. The occurrence of ascarid, intestinal impaction is very rare.
ii. Advice to the owner is:
- Treat any grazing cohorts of similar age with anthelmintic.
- Implement a parasite prophylaxis programme or modify the existing programme.
- Treat any grazing cohorts of similar age with anthelmintic.
Parasitological features relevant to control of ascarids are:
- Adult worms are prolific egg producers.
- Ascarid eggs are thick-shelled and resilient in the environment.

Although there is a potential risk of inducing further cases of ascarid impaction following treatment and death of large numbers of luminal worms, the probability of this is low. The drug of choice is either oral ivermectin (0.2mg/kg bodyweight) or oral fenbendazole (10mg/kg on five consecutive days). It will be necessary to repeat treatment at intervals of approximately 30 days as neither treatment regimen has high efficacy against immature ascarids early in migration. If possible, following treatment the group of animals should be grazed on pasture not used for young horses within the previous 12 months. As a result, in areas grazed by young animals with an inappropriate parasite prophylaxis programme it is likely that there will be a high level of transmission of parasites between horses.

Ascarid prophylaxis is best achieved by interval dosing all animals from three months of age with either pyrantel (four weekly), benzimidazoles (six weekly) or ivermectin (6–8 weekly). Alternatively, control could be achieved with daily in-feed dosing with pyrantel but this product is not licensed in Europe. Ideally, owners and managers should avoid using the same grazing area year-on-year for their young, growing animals.

125 & 126: Questions

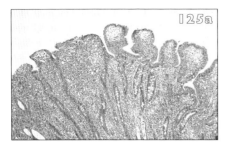
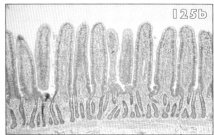

125 Shown in **125a** is a jejunal biopsy from a horse affected by chronic weight loss and small intestinal malabsorption (identified by oral glucose tolerance test). **125b** is a jejunal biopsy (from the same position in the bowel) from a normal horse for comparison.
i. What pathological abnormalities can you identify in **125a**?
ii. If the cellular infiltrate is composed predominantly of small lymphocytes and macrophages, what is the diagnosis?
iii. Which human disease has similar histopathological changes?
iv. Is this cellular infiltrate likely to be present in any other organs?
v. What is the aetiology of the equine disease?

126 A 12-year-old mare is presented to you with a history of intermittent unilateral haemorrhagic nasal discharge. The discharge occurs during exercise, but also at rest. During exercise a stertorous respiratory noise is audible. The mare is clinically normal, although airflow through the affected nostril is reduced compared to the other side. Endoscopy reveals a mass extending from the ethmoid region (**126**).

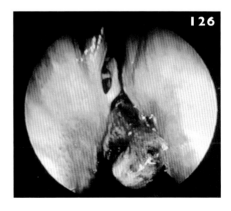

i. What is the likely diagnosis?
ii. What techniques can be used to confirm the diagnosis?
iii. What is the differential diagnosis of intermittent haemorrhagic nasal discharge?
iv. How would you treat this animal?
v. What is the prognosis without treatment?

125 & 126: Answers

125 i. There is villous atrophy. The villous height is dramatically reduced, and individual villi are atrophic and club-shaped. The lamina propria is infiltrated by a dense cellular infiltrate.
ii. Granulomatous enteritis.
iii. Crohn's disease.
iv. Yes. The mesenteric lymph nodes.
v. The precise aetiology of granulomatous enteritis is unknown. Immune-mediated responses to dietary, parasitic or bacterial antigens are probably involved. The disease has also been linked, in a few cases, to *Mycobacterium*, *Histoplasma* and *Campylobacter* infections.

126 i. Progressive ethmoidal haematoma (PEH).
ii. Endoscopy of the ethmoidal region may, as in this case, reveal an abnormal mass. The absence of a mass in the nasal passage does not exclude a diagnosis of PEH because small lesions can remain within the paranasal sinuses. Even when there is no apparent abnormality of the ethmoturbinates, the region should be carefully examined for traces of blood that might indicate the presence of small haematomas. PEH sometimes occurs bilaterally, so endoscopy must be performed bilaterally. Radiography and CT scan are indicated when there are no abnormalities visible during endoscopy. These techniques can also be used to provide the surgeon with information about the size, position and extent of the haematoma. Biopsy of a visible mass may confirm the diagnosis, but may result in considerable haemorrhage from the lesion.
iii. PEH; exercise-induced pulmonary haemorrhage (EIPH); trauma; nasal polyp; neoplasia (eg. fibrosarcoma); guttural pouch haemorrhage; fungal infections of the nasal cavity or paranasal sinuses; foreign bodies; clotting or bleeding disorders.
iv. Surgical treatment is indicated. Small and/or pedunculated lesions can be removed with Nd:YAG laser in the standing, sedated horse. Use of a wire snare or wire electrocauterization is also possible, but there is a higher recurrence rate. Larger lesions and those that originate from the sinuses must be treated by way of frontonasal or maxillary flaps. Combination of this technique with cryosurgery reduces the risk of haemorrhage during surgery. Intralesional injections with formalin have also been used successfully.
v. Prognosis without treatment is grave. Haematomas that appear in the nasal passages will eventually block the airway completely, causing suffocation. Haematomas in the sinuses cause progressive facial deformation and obstruction of the nasal cavity on the affected side.

127–129: Questions

127 You are called to examine an adult horse at a public boarding stable that has had an acute onset of diarrhoea (**127**). Two other horses recently have had similar clinical signs. You suspect an infectious and contagious aetiology.

i. What is the most common cause of acute infectious and contagious colitis in adult horses?
ii. What steps should be taken for initial management of this case?
iii. What procedures should be taken for control and prevention of further spread?

128 This horse lives on the coast of South Carolina. A six-year-old Thoroughbred brood mare, she presented with a one-month history of intermittent diarrhoea and weight loss (**128**). On physical examination, she had clinical signs consistent with endotoxaemia and low grade gastrointestinal pain. Upon gastrointestinal auscultation, a

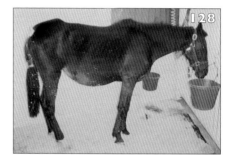

rushing sound was heard periodically in both ventral abdominal quadrants. Palpation per rectum was unremarkable, with the exception of the presence of gritty, poorly formed faeces in the rectum. You float some faeces in a rectal sleeve and a large quantity of sand settles to the bottom.
i. Why does this horse show signs of endotoxaemia?
ii. Is an abdominocentesis warranted in this case?

129 A 24-year-old retired Thoroughbred racehorse had exhibited repeated bouts of colic; euthanasia was carried out at the owner's request. At post-mortem examination the only grossly apparent abnormality was that illustrated in **129**. What is the significance of this finding?

103

127–129: Answers

127 i. Salmonellosis. (Potomac horse fever, an infectious but not contagious disease, should also be considered in endemic areas.)
ii. Implement a plan for fluid therapy (replacement and maintenance requirements), collect faeces for microbiological culture, and take blood for a CBC and serum biochemistry panel to evaluate neutrophil count, leukocyte morphology (evidence of endotoxaemia and/or septicaemia), protein concentration, azotaemia, and electrolyte and albumin concentrations in serum. If in an endemic area, testing for *Ehrlichia risticii* may be indicated. Results will guide therapy and provide information for determining if adjunct treatment with plasma, antimicrobials or other agents to combat septicaemia and/or endotoxaemia (eg. low-dose flunixin meglumine) is indicated.
iii. Principles of control and prevention of salmonellosis have been reviewed recently (Newton-Clarke M: *Equine Vet Educ* 1995;7:67–69). Briefly, prevention efforts should focus on eliminating contamination of the environment (isolation of affected animals; protective clothing such as disposable gowns, boot covers, gloves, masks, etc.; minimize fomite transmission by not carrying stall cleaning utensils, buckets, brushes, etc. from horse-to-horse), and cleaning and disinfecting contaminated or suspect areas. Cleaning (preferably with steam) should be followed by disinfection (preferably with a phenolic compound because of activity in the presence of organic material). A 10% bleach solution can be an effective disinfectant if organic material is not present.

128 i. Sand is extremely irritating and can abrade the colonic mucosa, allowing translocation of endotoxin from the gastrointestinal lumen into the peritoneal cavity and from there into the general circulation.
ii. No. If the clinical diagnosis has been made, then an abdominocentesis should not be performed unless the horse's condition changes or deteriorates. The large amount of sand that may be present in the colon greatly increases the chances of enterocentesis, as the heavy colon lies directly on the floor of the abdomen, against the peritoneal lining.

129 Infestation with the tapeworm *Anoplocephala perfoliata* occurs in about 70% of equine animals worldwide and is sometimes associated with mucosal erosions and/or inflammation at sites of parasite attachment near to the ileocaecal junction. From epidemiological studies of risk factors of colic, there is some preliminary evidence that *A. perfoliata* may be associated with specific ileal or caecal disorders, including intussusceptions and ruptures of either of those intestinal areas. The fact that these are rare disorders, yet the prevalence of tapeworm infection is high, would suggest that the pathogenicity of the worms should not be overstated. The only drugs licensed for use in the horse with proven high efficacy against tapeworms are pyrantel salts at double the dose rate recommended for nematodes. Praziquantel is not licensed for the horse but has been found to have 85% and 95% efficacy against tapeworms at dose rates of 0.5mg/kg and 1.0mg/kg respectively.

130 & 131: Questions

130 A nine-week-old Arabian foal is presented with a history of recurrent respiratory infections characterized by nasal discharge, cough, dyspnoea and pyrexia. The foal was normal at birth, but suffered its first bout of respiratory disease at six weeks of age. Over the previous week it had also developed widespread scaling and alopecic skin lesions (**130**), from which

Dermatophilus congolensis had been cultured. The foal has also had intermittent diarrhoea. Haematological evaluation revealed a severe lymphopenia (lymphocyte count 0.2×10^9/l).
i. What is the most likely diagnosis?
ii. How could you confirm this?
iii. What is the aetiology of this disease?
iv. What are the usual causes of the respiratory infections in this disease?
v. How would you treat this foal?

131 A 10-week-old foal in good body condition showed signs of acute dyspnoea.
i. Describe the radiographic findings illustrated in **131**.
ii. What is the most likely aetiologic agent?
iii. What is the source of the infection for the foal?
iv. What is the recommended treatment?

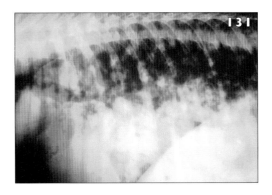

130 & 131: Answers

130 i. Combined immunodeficiency syndrome (CID).
ii. Serum IgM will be extremely low or undetectable. The lymphopenia will be persistent. Lymphoid hypoplasia, including the thymus, spleen and lymph nodes, will be evident at post-mortem examination.
iii. CID is a genetic disorder of foals of Arabian breeding that results in failure to produce functional B and T lymphocytes. The disease is inherited as an autosomal recessive trait, and heterozygous individuals show no signs. Affected foals are usually normal at birth, and signs of infection (especially respiratory infections and pneumonia) become evident at any time between two days and two months depending on the level of passively acquired colostral immunity.
iv. Pneumonia caused by adenovirus, *Pneumocystis carinii* and a variety of bacteria are commonly involved.
v. There is no recommended treatment for CID. Bone marrow transplantation has been performed successfully in one affected foal.

131 i. Lateral view of the thorax demonstrates nodular lesions in the cranioventral lung fields. These lesions, some of which are cavitary, obscure the cardiac and caudal vena cava silhouette. Tracheobronchial lymphadenopathy is suggested by the elevation of the trachea.
ii. Radiographic lesions of this type seen in a 10-week-old foal are almost pathognomonic for *Rhodococcus equi* infection.
iii. *Rhod. equi* is a Gram-positive, pleomorphic, facultative, intracellular, obligate aerobic bacteria. *Rhod. equi* lives in faeces and soil and is resistant to most chemical and environmental conditions. The organism is found in the intestines of many normal mammals including horses. Foals are most often infected by inhalation of the organism from the soil or faeces of the animals, but ingestion or umbilical entry are also possible.
iv. Treatment of *Rhod. equi* has been difficult due to the intracellular characteristics of the organism. The combination of erythromycin and rifampin is excellent *in vitro* and has decreased the mortality rate associated with *Rhod. equi* pneumonia. Erythromycin is a non-polar, macrolide antibiotic that penetrates caseous material readily. Several oral forms have been available for use in horses. Erythromycin causes gastrointestinal irritation, which may cause mild diarrhoea in horses. Erythromycin phosphate or stearate is dosed at 37.5mg/kg p/o bid, while erythromycin estolate is used at 25mg/kg p/o tid or qid. Rifampin is synergistic with erythromycin and penetrates macrophages, neutrophils and caseous material readily. Rifampin is dosed at 5mg/kg p/o bid, although some clinicians use 10mg/kg p/o sid.

132 & 133: Questions

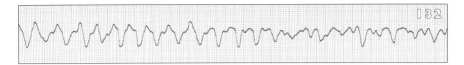

132 A three-day-old foal is being treated for septicaemia. Suddenly, it loses consciousness and stops breathing. The mucous membranes are pale and there is no palpable pulse.
i. What does the ECG show (**132**)?
ii. How should this be treated?

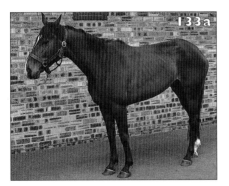
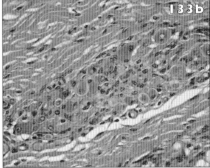

133 A seven-year-old hunter mare at pasture for the summer had a history of marked weight loss, intermittent mild colic and occasional oesophageal spasm for six days (**133a**). No food and little water had been consumed and only a few hard faecal pellets had been passed over that time period. The results of a CBC, serum proteins and blood urea were consistent only with dehydration. No worm eggs were present in the faeces, and liver enzymes, plasma fibrinogen and peritoneal fluid analysis were normal. An exploratory laparotomy was undertaken. The whole gastrointestinal tract was abnormally empty of ingesta and apparently incoordinated peristaltic movements of the small intestine were observed. A 20 x 10mm full thickness biopsy of the ileum was taken for histological examination (stained with haematoxylin and eosin, x25, **133b**).
i. Describe the histopathological features present.
ii. If this pony is from Scotland, England, Wales, mainland Europe or Argentina, what is your diagnosis?
iii. What other histopathological changes occur in this disease and in which tissues?

132 & 133: Answers

132 i. The ECG shows ventricular fibrillation characterized by random undulations due to chaotic electrical activity.
ii. The priorities for treatment are:
A – AIRWAY – insert an orotracheal or nasotracheal tube.
B – BREATHING – initiate mechanical ventilation. In foals a resuscitator bag (Ambu) designed for humans is ideal. Ventilate at around 30bpm.
C – CIRCULATION – establish intravenous access and administer polyionic fluids rapidly. Electrical defibrillation is the treatment of choice for ventricular fibrillation. In the absence of this, external cardiac massage is initiated; adrenaline, intravenously or intratracheally, increases vascular tone and makes cardiac massage more effective; lignocaine (lidocaine) and/or bretyllium may reverse the arrhythmia.

133 i. The section shows a group of neurons which constitute part of the myenteric plexus which lies between the inner circular and outer longitudinal muscular layers. The neurons show loss of Nissl substance (basophilic rough endoplasmic reticulum with a granular appearance) so that the cytoplasm is homogeneously eosinophilic, and pyknosis, margination or loss of nuclei. **133c** shows the myenteric plexus from the ileum of a normal horse for comparison. The basophilic Nissl substance in the cytoplasm and normal 'clockface' nuclei are visible (stained with haematoxylin and eosin, x25).
ii. The degenerative changes observed are pathognomonic of grass sickness (equine dysautonomia). In this animal the duration of clinical signs (between three and seven days) would classify the case as subacute grass sickness.
iii. Neuronal degeneration also occurs in the pre- and paravertebral autonomic ganglia (eg. anterior cervical, stellate, intervertebral and coeliacomesenteric ganglia), certain brainstem nuclei (most consistently the oculomotor but also the trigeminal, hypoglossal, lateral vestibular and facial nuclei and the dorsal motor nucleus of the vagus) and in the intermediolateral nucleus of the spinal cord. Within the gastrointestinal tract, the lesions are most severe in the ileum with the oesophagus, duodenum, jejunum and large intestine showing less severe degenerative neuronal changes. The rectum is inconsistently affected so that rectal biopsy is not a recommended diagnostic test. The myenteric plexus is shown in **133b** but the submucous plexus is also affected.

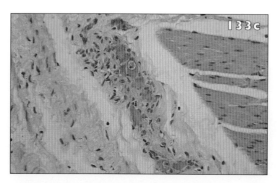

134–136: Questions

134 You are asked to examine a three-year-old pony gelding because it has been sluggish for three months. Clinical evaluation reveals a high pulse rate (60/minute) in combination with mydriasis.
i. What is the most likely endocrinopathy?
ii. Which other clinical signs besides tachycardia and bilateral mydriasis can you expect?
iii. Describe normal catecholamine production in the horse?
iv. Is an abdominocentesis indicated?
v. How would you confirm the diagnosis?
vi. How would you treat this horse?

135 A 12-year-old pony gelding is presented with a three-week history of weight loss, inappetence, intermittent fever (up to 38.7°C) and ventral thoracic and abdominal oedema (**135**). Auscultation of the chest reveals bilateral absence of ventral lung sounds, and a pleural effusion is confirmed by ultrasonography. Thoracocentesis yields a large volume of blood-stained watery fluid. A tracheal wash did not reveal any any infectious cause.

i. What is the most likely diagnosis?
ii. How can the diagnosis be confirmed?

136 This horse's spleen (**136**) shows three distinctly depressed clefts in its upper half, parietal surface near the anterior border, and a fourth one near the middle of the upper parietal surface. The three clefts near the anterior edge have a roughly ragged edge. What are these four clefts called, and what is their significance?

134–136: Answers

134 i. The most likely endocrinopathy is phaeochromocytoma. Phaeochromocytoma is rare but the most often diagnosed adrenal medullary neoplasm of horses. Although both adrenal glands may be affected, phaeochromocytoma is usually unilateral. One case has been described in a six-month-old Standardbred filly with spinal metastases.
ii. Clinical signs of phaeochromocytoma are largely attributable to the increased production and release of catecholamines. These clinical characteristics are sweating, tachypnoea, muscle tremors, hyperglycaemia and colic.
iii. In the horse it is not certain whether epinephrine or norepinephrine is predominantly produced by phaeochromocytomas. About 20% of the adrenal catecholamines in the normal horse is norepinephrine. The catecholamines are synthesized in the chromaffin cells of the adrenal medulla. However, most of the plasma norepinephrine is derived from sympathetic nerves. In addition, there is evidence that both norepinephrine and epinephrine are released by neural tissue adjacent to the hypothalamus.
iv. Yes. Phaeochromocytomas are predisposed to haemorrhage, and this condition should be considered in cases of colic associated with blood in the abdominal cavity.
v. Diagnosis must be based on measurement of circulating catecholamines and/or their urinary metabolites, eventually combined with blood pressure measurements. However, assays measuring catecholamines are not readily available.
vi. Although (bilateral) adrenalectomy can be performed via (bilateral) transcostal retroperitoneal approaches through the 18th ribs, therapeutic intervention often turns out to be too late. Prior to the surgical intervention the animal has to be checked for metastases. The bilateral adrenalectomized horse must be supplemented with up to 75mg of desoxycorticosterone pivilate i/m every 2–4 weeks, and up to 50mg of prednisolone orally once a day.

135 i. Thoracic neoplasia. The commonest type of thoracic neoplasia in the horse is mediastinal lymphosarcoma.
ii. Cytology of pleural fluid may reveal neoplastic cells. Similar fluid may be present in the peritoneal cavity. Neoplastic masses may be identifiable by thoracic radiography or ultrasonography. Mediastinal masses may sometimes extend through the thoracic inlet and be palpable externally at the base of the jugular groove; in this situation they may be amenable to biopsy. In the case of lymphosarcoma, peripheral lymph nodes may be enlarged due to neoplastic infiltration, and these may be biopsied.

136 These are normal structures commonly seen in horses of all ages. They are congenital and are called capsular splenic folds. They are of no clinical significance except that they are commonly misdiagnosed as pathological lesions.

137–139: Questions

137 The foal pictured in **137** was presented for evaluation of colic. It is from an Overo Paint horse mating. What is the likely diagnosis, the prognosis and the responsible lesion?

138 A client has to move her three horses, aged two, three and 16 years, to a new property where a case of grass sickness occurred the previous year. She is concerned about the risk to her own horses. An outline plan of the property is shown in **138**.
i. What are the major epidemiological features of grass sickness?
ii. How can the risk of the client's horse developing grass sickness be minimised, bearing in mind the layout of the property?

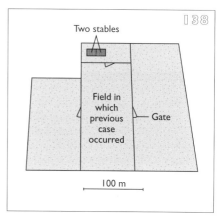

139 A three-year-old horse suddenly developed a crop of nodules in its skin in October (**139**). The owner reports that the horse is somewhat lethargic and takes longer to eat its grain than usual. On palpation, the nodules are very tender and somewhat soft to the touch. One of the nodules is aspirated and only a small amount of material is withdrawn.
i. List the two most likely differential diagnoses.
ii. Describe the cytological findings for each of the above differentials.

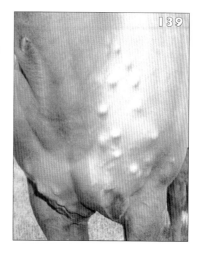

137–139: Answers

137 The foal has ileocolonic aganglionosis ('lethal white' foal). The prognosis is grave, and euthanasia would be expected to be performed. The lesion is a congenital aplasia of the mesenteric ganglia.

138 i. Grass sickness occurs in Europe and the southern tip of South America. The highest incidence is in the UK, especially the east of Scotland where 1% of horses die annually from the disease in some areas. All Equidae are susceptible and almost all affected animals are at pasture for at least part of the day. The aetiology is unknown but a mycotoxin is suspected. Young animals (2–7 years) are predisposed, as are horses which have recently moved pastures. The peak seasonal incidence is in the spring and summer with a peak in May in the northern hemisphere. Dry, cool weather often occurs for 10–14 days prior to the occurrence of cases. It is not contagious but certain premises or specific fields will repeatedly produce cases. Sporadic cases and case clusters can occur.

ii. Since the horses are being moved to a premises where grass sickness has recently occurred, they must be regarded as having a high risk of contracting the disease. They are at greatest risk for the first two months on the new property and it would be advisable to stable them, allowing no access to pasture for this two month period. They can then be gradually introduced to grazing. There is limited stabling available, so the two young horses should have priority as they are at much greater risk than the 16 year old. As it is known in which field the previous case occurred, the use of this field for grazing horses should be avoided if possible, especially during the season of highest risk. However, it is still possible for cases to occur in the adjacent fields. Especially during high risk periods, the horses should not be allowed to graze the affected field while crossing it to reach the other fields as the disease can occur in horses with only a few minutes daily access to affected pasture. Once the horses are at grass, stabling is still advisable if more than 7–10 consecutive days of dry weather occur, with a minimum daily temperature of 7–11°C, as such weather conditions often precede the occurrence of cases. Unfortunately, no preventive measures other than permanent stabling will eliminate the chance of grass sickness occurring.

139 i. Infectious furunculosis, nodular panniculitis (rare).

ii. The cytological findings for each of the above are:
- *Infectious furunculosis*: pyogranulomatous inflammation with toxic neutrophils. Depending on which agent is causing the infection, it may or may not be seen.
- *Nodular panniculitis*: pyogranulomatous inflammation. Very few neutrophils will show degenerative changes and the macrophages will contain numerous cytoplasmic fat-filled vacuoles.

140–142: Questions

140 Auscultation of the ventral lung fields of a four-year-old Thoroughbred-cross gelding revealed a decrease in breath sounds.
i. What diagnostic procedure is illustrated in 140?
ii. What is the value of this procedure?
iii. What are the normal values for pleural fluid nucleated cell counts and total protein?

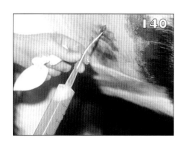

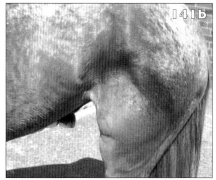

141 The horse illustrated (**141a, 141b**) is a two-year-old Thoroughbred-cross filly. She has been presented for investigation of progressive weight loss over a period of three weeks during which time her appetite has been capricious and four episodes of mild colic have occurred. Clinical examination revealed a heart rate of 60bpm at rest. Fine muscle tremors were noted in the triceps region and areas of patchy sweating were present on the neck and flank. The filly lives in west Wales and is pastured during the daytime. What disease would be top of your list of differential diagnoses, and how would you set about confirming or refuting your suspicions?

142 This stone has just been removed from the bladder of a race mare without performing a laparotomy or any surgical incision (**142**).
i. How was the stone removed?
ii. What is the most likely chemical composition of the stone?
iii. What was the most likely clinical complaint?

140–142: Answers

140 i. Thoracocentesis, a surgical puncture of the chest wall into the pleural cavity for aspiration of fluids, also called pleurocentesis.
ii. Thoracocentesis is a procedure which has both diagnostic and therapeutic value. When the findings from either thoracic auscultation or percussion suggest a pleural effusion, thoracocentesis can both confirm its presence and provide a specimen for examination. Analysis of the pleural fluid may in turn help you to determine the underlying disease process and develop a therapeutic plan. Drainage of pleural effusion via thoracocentesis is beneficial in removing large volumes of fluid from the thorax. Immediately following this drainage, some horses show increased pain, associated with loss of the 'cushion' of fluid in the pleural space. Removal of this fluid decreases respiratory effort and is beneficial in the resolution of an infectious process in the thorax. In horses with thoracic neoplasia and large volumes of pleural effusion, thoracocentesis and chest drainage may resolve signs of dyspnoea and respiratory distress.
iii. Pleural fluid from healthy horses may contain up to 10,000 nucleated cells/µl and 35g/l total protein. Most horses have less than 5,000 nucleated cells/µl and less than 25g/l total protein.

141 This horse is suffering from grass sickness (equine dysautonomia) in its chronic form. The history and clinical signs are typical of this condition. Other historical and epidemiological associations with this condition include: previous history of grass sickness on premises; young animal, recently moved to new pasture/new stables; spring and autumn peak incidence; reduced faecal output; diarrhoea; 'quidding' of food; intermittent hypersalivation. Ante-mortem confirmation of a clinical diagnosis can be difficult. The identification of defects in oesophageal function by means of barium contrast radiography can be a useful ancillary test in some cases. However, interpretation of such radiographs is subjective and the test lacks specificity for grass sickness.

The only ante-mortem test currently available to confirm this diagnosis is that of ileal biopsy. The histological changes that characterize the condition are neuronal degeneration and necrosis. Such changes occur in the autonomic ganglia and in the enteric nervous system. An elliptical biopsy 1–2cm in length should be taken from the anti-mesenteric border of the ileum at the level of the termination of the ileocaecal ligament. Formalin fixed biopsies should be sent to a pathologist familiar with the technique. A result can usually be obtained within 24–48 hours of the specimen arriving at the laboratory.

142 i. This stone was removed per vagina after xylazine epidural.
ii. The most frequent predominant composition of urinary calculi in adult horses is calcium carbonate.
iii. Haematuria associated with exercise is the most common complaint with cystic calculi in performance horses. Dysuria may be pronounced if the stone is partially obstructing urine outflow from the bladder.

143 & 144: Questions

143 A 12-year-old Welsh Cob maiden mare is presented in May with a history of being scanned pregnant by another practice last autumn, four months after breeding but did not gain weight and was clearly not pregnant in late winter. She has not been seen in oestrus this spring.

On palpation a large, fluid filled structure can be balloted in the abdomen. A small,

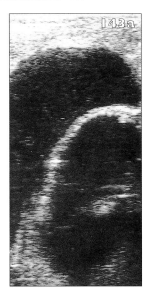
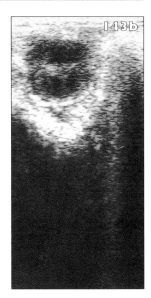

featureless right ovary can be palpated. The right uterine horn feels normal, but although the base of the left horn can be palpated, it seems to merge with the large smooth palpable structure. The left ovary cannot be identified. **143a** and **143b** show the ultrasound scan pictures of the mass on the left and the right ovary, respectively.
i. Based on your findings, what are the differential diagnoses and what is the relative incidence of each option.
ii. What further diagnostic tests could be performed?

144 You are asked to examine a five-year-old pony with a two-day history of dullness, inappetence, reluctance to move, tachycardia (heart rate 60bpm), tachypnoea (respiratory rate 18bpm) and mild abdominal pain. The rectal temperature is 38.6°C. You perform abdominocentesis (**144**).
i. What is the most likely diagnosis?
ii. What is the aetiology and pathogenesis of this disease?

143 & 144: Answers

143 i. 143a and **143b** suggest that the diagnosis of pregnancy was incorrect. It seems likely that the abdominal fluid-filled mass was mistaken for a pregnant uterus. The mass could be uterine obscuring the ovary, or ovarian and adherent to the uterus. The incidence of uterine masses is very low. Uterine tumours are rare, as are uterine haematomas or abscesses. It is unlikely that a uterine mass would interfere with ovarian cyclicity. The fact that the mare does not appear to be cycling implies that the mass is hormone-secreting and therefore likely to be ovarian in origin. The difficulty experienced in distinguishing uterine horn from ovary suggests that the ovary is adherent to the horn. By far the commonest hormone-secreting ovarian tumour is a granulosa cell tumour. These tumours can secrete testosterone or oestrogen and also inhibin. It is the negative feedback of these hormones on the hypothalamus and pituitary which causes the contralateral ovary to atrophy. However, in occasional mares the contralateral ovary will still be functional and the mare will cycle normally. Other differentials for an enlarged ovary include haematoma (the commonest cause for ovarian enlargement in the mare), pregnancy (days 40–150 when supplementary corpora lutea are present) and persistent anovulatory follicles. Each of these conditions, however, has a characteristic ultrasonographic appearance, is not associated with behavioural abnormalities, and haematomas and anovulatory follicles will eventually diminish in size after several weeks.

ii. Further diagnostic tests that could be performed are:
- *Blood hormone levels.* Elevated concentrations of serum testosterone in a mare are diagnostic for a granulosa cell tumour. However, only about 50% of affected mares have elevated levels of this hormone. Mares with high testosterone tend to display aggressive, stallion-type behaviour. Approximately 87% of mares with granulosa cell tumour have elevated circulating concentrations of inhibin. Unfortunately, assays for inhibin are not available commercially in the UK at this time.
- *Uterine endoscopy.* Endoscopy revealed a normal right horn but the left side appeared blind-ending, one third of the way up the horn.

The ultrasonographic appearance of granulosa cell tumours is too variable to allow definitive diagnosis.

144 i. Peritonitis (septic peritonitis).
ii. There are a large number of potential causes of septic peritonitis. Peritonitis is inflammation of the mesothelial lining of the peritoneal cavity. Septic peritonitis is caused by bacterial infection of the peritoneal cavity. Potential causes include gastrointestinal perforation, intestinal infarction and ulceration, abdominal abscessation, complications of castration and abdominal surgery, enterocentesis, penetrating abdominal wounds, genitourinary tract perforations and cholelithiasis.

145–147: Questions

145 You are asked to examine an eight-year-old Quarterhorse gelding. The owners found the horse with feed coming from its nostrils, stretching its neck and attempting to drink without apparent success. The horse is fed a diet of grass hay and sweet feed. Although the owners have only owned the horse for two months, the gelding is considered a greedy eater. The children of the owners and their neighbours are quite fond of the horse, and the

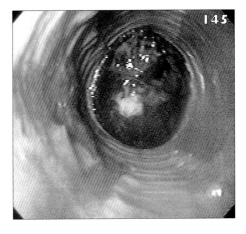

children often offer the horse assorted 'treats'. Oral examination reveals very sharp enamel points and large dental hooks on the upper 2nd premolars and lower molars. You attempt to pass a stomach tube, but you encounter resistance at the distal region of the cervical oesophagus. Endoscopic examination reveals the image shown in **145**.
i. What is your diagnosis?
ii. What do you recommend for treatment?
iii. What do you recommend for prevention?

146 i. What is the aetiology of guttural pouch haemorrhage?
ii. What complications may develop from this disease?

147 A 10-year-old Appaloosa mare has had a chronic ocular discharge for several months from this eye (**147**). The owner thinks that the eye looks redder than normal, but the horse squints when the owner tries to get a close look at the eye. What is the most likely diagnosis?

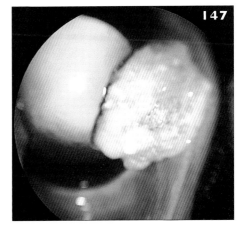

145–147: Answers

145 i. Oesophageal obstruction with a potato.
ii. Sedation, placement of a nasotracheal tube (if available) and lavage with warm water (lubricants such as mineral oil may help but will be worse if aspirated). It may be necessary to use general anaesthesia to enable relief of the problem. After relieving the obstruction, a soft diet (mash or gruel of pellets) should be fed for at least a few days because the mucosal damage is likely to have been circumferential. Occasionally, it may be advisable to withhold feed for 24–48 hours before starting the horse back on a soft diet. Broad-spectrum antimicrobials may be indicated to treat or prevent bacterial pneumonia associated with aspiration, and non-steroidal anti-inflammatory agents may be administered to reduce swelling and pain. The horse should have access to plentiful fresh water. Because of the potential for stricture formation secondary to circumferential mucosal damage, the horse should be re-evaluated endoscopically approximately 2–4 weeks after relieving the obstruction, because this is when the oesophagus will be most narrow; at 60 days, the oesophagus will have achieved maximal diameter.
iii. Prevention should include instructing children not to offer this horse treats like apples and potatoes, providing dental care to the horse, maintaining a soft diet and using methods to slow his consumption, such as placing rocks in his feeder.

146 i. Guttural pouch haemorrhage is most commonly a complication of guttural pouch mycosis. The mycosis usually affects the dorsal wall of the medial compartment of the pouch, and a diphtheritic membrane composed of necrotic tissue develops at this site. Fungal mycelia are present throughout the diphtheritic membrane, and also invade the underlying tissues. Haemorrhage arises from erosion of the internal carotid artery or, less commonly, the external carotid or maxillary arteries. A number of fungal species can be found in these lesions, including *Aspergillus fumigatus* and *A. nidulans*, but their aetiological role is unclear since these species are commonly cultured from the upper airways of healthy horses. As in other fungal diseases, a disturbance of host defence mechanisms may enhance the ability of fungi to cause disease.
ii. Without treatment, the animal will probably die due to extensive blood loss from consecutive haemorrhages. Sometimes, the initial haemorrhage will be fatal. Damage to cranial nerves (eg. the vagus and glossopharyngeal nerves) which lie adjacent to the guttural pouch, can result in dysphagia. Horner's syndrome and facial nerve paralysis have also been reported.

147 Corneal–conjunctival squamous cell carcinoma. The temporal limbal region is the most common location for squamous cell carcinoma of the globe. However, the lesion may originate from any conjunctival site and progress across the limbus.

148 & 149: Questions

148 An eight-year-old gelding is presented with acute colic, abdominal distension, trembling, sweating, tachycardia (heart rate 76bpm) and tachypnoea (respiratory rate 32bpm). The horse had accidentally eaten a large quantity of grain several hours earlier. The mucous membranes are bright red with petechial haemorrhages (**148**). Intestinal sounds are absent, and percussion of the abdomen results in high-pitched pings. Colonic distension with tight bands are palpated *per rectum*. Haematology reveals haemoconcentration (PCV, 0.68l/l), neutropenia (PMN count, 3.7×10^9/l) with toxic changes in the neutrophils.

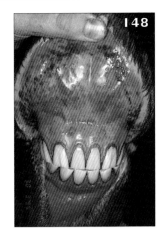

i. What is the diagnosis and pathogenesis?
ii. How would you treat this horse?
iii. What is the prognosis?

149 An eight-year-old Throughbred mare is being transported to your clinic because of two days of progressive weakness and ataxia involving all four limbs (**149a**). The ataxia and weakness have been noticeably worse in the hindlimbs and dysuria has been consistently noted during the course of the disease. The mare has remained bright and alert with normal appetite. Two other horses at the same riding stable were similarly affected. Both became recumbent within 24 hours of onset of clinical signs and euthanasia was necessary. On presentation this mare is also recumbent and cannot stand. All cranial nerve function appears normal. A lumbosacral tap is performed and the fluid appears grossly discoloured (**149b**), has a high protein (157g/l) but only one WBC/µl.

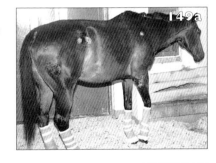

i. What is the diagnosis?
ii. Why is the spinal tap discoloured?
iii. What are the landmarks for performing the lumbosacral spinal tap?
iv. What would the appropriate medical treatment be for this horse?

148 & 149: Answers

148 i. Grain overload. Grain engorgement results in rapid fermentation and the production of large amounts of lactic acid, volatile fatty acids and gas in the gastrointestinal tract, leading to gastrointestinal atony and distension. The severe acidosis results in massive endotoxaemia and subsequent circulatory failure (shock). The presence of petechial haemorrhages is suggestive of disseminated intravascular coagulation.
ii. Intravenous fluid therapy is required to combat circulatory failure. Hypertonic saline may be administered initially, but this must be followed by polyionic fluid (2–4 litres/hour). Plasma may also be beneficial (2–4 litres), preferably hyperimmune plasma containing antibodies against endotoxins. Flunixin meglumine (0.25mg/kg every 8 hours) is given after signs of colic have abated for its anti-endotoxic effects. A nasogastric tube should be passed, and may be left in place to relieve gastric distension. If there is no gastric reflux, activated charcoal may be administered in water via the nasogastric tube. The horse should not be allowed to eat until the clinical signs have improved. Frog pads may be applied to the front feet in an attempt to prevent or minimize the risk of laminitis.
iii. The prognosis for horses with severe clinical signs such as this is poor. The prognosis is particularly poor in cases with severe abdominal pain and abdominal distension, and horses which develop early signs of laminitis.

149 i. Equine herpesvirus 1 myeloencephalopathy.
ii. The spinal fluid has a yellowish discolouration because of prior leakage of red blood cells into the spinal fluid. The clinical signs and CSF changes are a result of a vasculitis of the CNS.
iii. An 18G, 15cm spinal needle is used to collect the fluid. An imaginary line is drawn across the dorsum of the horse connecting the right and left caudal borders of the tuber coxae. The site for introduction of the needle is just caudal (1.25–2.5cm) to this imaginary line. The cranial edge of each tuber sacrale may be palpated on either side of this site in some thin horses. The tap is performed on the midline of the horse and the needle is introduced perpendicular to a depth of 11.25–15cm.
iv. The mare should be treated with corticosteroids (dexamethasone 0.06–0.4mg/kg) because of the rapid progression of the disease. Bactericidal antibiotics should be administered to help prevent bacterial pneumonia and cystitis. The bladder should be catheterized as needed to prevent prolonged distension. In the case described here, the mare recovered with this treatment.

150 & 151: Questions

150 A year-old Miniature Shetland colt has exhibited signs of mild colic on three occasions during a 14 day period. There are a total of 15 ponies of various ages kept on the same premises and the owner reports that at least six other individuals have had similar colic signs within the previous year. On physical examination the pony is in moderate body condition but no other abnormal physical findings are identified. On blood biochemical analyses there is a moderate hypoalbuminaemia (24g/l) and elevated alkaline phosphatase (496u/l). No haematological or peritoneal fluid cytological abnormalities are present but faecal parasitological examination revealed 550 strongyle eggs per gram (epg) and also 1,250 *Parascaris equorum* epg. Given the history of multiple clinical cases of colic as well as the clinical and clinicopathological findings:
i. What is the most likely diagnosis?
ii. How would you further investigate the case?

151 This four-year-old Quarterhorse mare (**151a, 151b**) developed a swelling over her right eye at 08.00 hours. Within hours the swelling involved the entire head and was progressing down the neck. Eight hours after the onset, a tracheostomy was performed because of severe upper respiratory compromise. A venomous bite is suspected, although no wound can be identified. Now the mare has begun to show signs of epistaxis, and she has developed a thrombosed cephalic vein at the site of catheter placement six hours previously.
i. What pathophysiological process do you suspect is responsible for the epistaxis and cephalic thrombosis?
ii. What diagnostic tests would you do to confirm your suspicion?

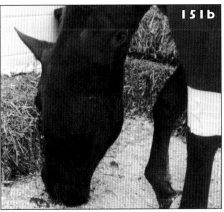

150 & 151: Answers

150 i. Strongyle-associated colic, but dietary causes in particular cannot be ruled out.
ii. (a) Establish specific details of the parasite prophylaxis programme practised for the whole group of animals.
(b) Perform faecal parasitology on the whole group of animals.
(c) Perform blood biochemistry and haematology on the whole group of animals.
(d) Perform pasture larval counts on grazing areas.
(e) Obtain an accurate profile of the clinical incidents to:
- Confirm that they can reasonably be considered to be the same condition.
- Identify risk factors such as age, specific paddock, transport, weaning, etc.
- Assess the effectiveness of any specific treatments.
(f) Objectively assess management practices.

With specific reference to parasite-associated disease, diagnoses are commonly ruled out inappropriately by focusing on the historical and/or clinical and/or clinico-pathological data of the individual case rather than the group of animals in which the animal is managed. In the case described the faecal parasitology counts indicate adult burdens of both strongyle and ascarid worms. *P. equorum* is a common parasite of young animals. It is rarely associated with colic and so is not likely to be a cause of multiple cases in a group and certainly not if adult animals are affected. The main relevance of parasitic infection in this case is as a possible contribution to the poor body condition and as an indicator that parasite control is suboptimal. The evidence of a mid-range strongyle faecal egg count is not enough by itself to confirm the diagnosis of strongyle-associated colic.

A circumstantial diagnosis of strongyle-associated colic can be made on the basis of either: faecal strongyle egg counts of over 400 strongyle epg in 25% of the group; and/or hypoalbuminaemia/hyper(beta)globulinaemia/neutrophilia in 25% of the group; and/or high pasture strongyle larval counts from the grazing area.

151 i. Disseminated intravascular coagulopathy (DIC) secondary to a massive, venom-induced inflammatory reaction.
ii. Platelet count, fibrin degradation products concentration, anti-thrombin III activity. These three tests are the most sensitive indicators of DIC available. Platelets are consumed in the unregulated activation of coagulation, fibrin degradation products are released as an activated fibrinolytic system attempts to restore homeostasis in the face of unregulated coagulation, and anti-thrombin III, the natural anti-coagulant, is consumed as it binds to activated thrombin in another attempt to reduce the rate of clot formation.

152–154: Questions

152 A five-year-old pony gelding is presented with acute onset of watery diarrhoea, depression, inappetence and pyrexia (39.7°C) (**152**).

i. Why are the terms 'colitis' and 'diarrhoea' often used synonomously in adult horses, but not in foals?
ii. What are the major causes of acute diarrhoea in adult horses?
iii. What are the major causes of acute diarrhoea in nursing foals?

153 A six-year-old Standardbred stallion has a recent history of poor training performance characterized by inability to pace a mile in less than two minutes twenty seconds, signs of mild abdominal discomfort and grunting. A greatly enlarged (~16cm in length) left kidney was palpated on rectal

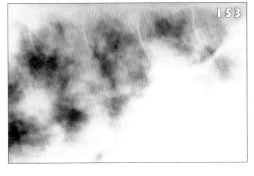

examination. Crackles and wheezes were auscultated over both lung fields and the horse coughed and grunted following deep inspiration.
i. What are the radiographic findings in **153**?
ii. If these masses are tumours, is primary or metastatic pulmonary neoplasia more likely?
iii. Given the rectal findings in this horse, what is the likely diagnosis?
iv. Name two diagnostic tests that could be used to determine the nature of the tumour?

154 i. Persistent seizures or a persistent maladjustment syndrome may be present in what clinical condition pictured in **154**?
ii. How can the diagnosis be confirmed if external signs are not present?

152–154: Answers

152 i. Diarrhoea in adult horses is virtually always the result of colonic disease or dysfunction, hence the term colitis. In foals, on the other hand, diarrhoea is most severe with small intestinal disease.
ii. Salmonellosis, Potomac horse fever (in endemic areas), larval cyathostomosis, antibiotic-associated diarrhoea, intestinal clostridiosis, colitis X, dietary causes, endotoxaemia.
iii. Foal heat diarrhoea, nutritional causes, rotavirus, salmonellosis, enterotoxigenic *Escherichia coli*, cryptosporidiosis, intestinal clostridiosis, necrotizing enterocolitis, *Strongyloides westeri*, antibiotic-associated diarrhoea.

153 i. Multiple patchy areas of increased density throughout the caudodorsal lung field.
ii. Primary pulmonary neoplasia is seen in the horse, but not as frequently as metastatic disease. The distribution of the lesions in the radiographs is more typical of metastatic lung disease.
iii. Metastatic renal adenocarcinoma
iv. Diagnostic tests to consider include biopsy of either the lung or enlarged kidney, or transtracheal aspirate, bronchoalveolar lavage or pleural fluid cytology. Frequently, metastatic pulmonary masses do not exfoliate into either the airways or the pleural space and examination of these specimens may be normal. Thoracic ultrasound may help identify a site for lung biopsy which can be done with local anaesthesia and little to no sedation.

154 i. The foal has external hydrocephalus. Usually the calvarium is soft and ultrasound of the cranium through the open fontanels can reveal the excessive cranial vault fluid.
ii. Internal hydrocephalus would not provide for a physical diagnosis, and would require a CT scan or magnetic resonance imaging (MRI). Electroencephalograms (EEG) may provide additional information.

155–157: Questions

155 A five-month-old Quarterhorse colt is examined because of progressive stiffness of the hindlimbs and difficulty in rising. On examination, both bulging and dimpling of the semimembranosus and semitendinosis are noted (155). These abnormalities are not always present and can be induced sometimes by firmly tapping the muscles with a finger.

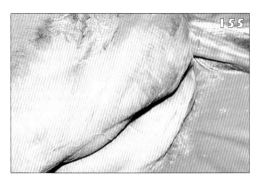

i. What is the diagnosis?
ii. What is the prognosis for performance?
iii. What would EMG testing reveal?

156 The mare described in 143 has a granulosa cell tumour of the left ovary that is adherent to the left uterine horn. What is the treatment and prognosis for future fertility?

157 A three-month-old Belgian foal (157) is examined for an acute onset of stupor, blindness, obsessive circling and/or head pressing. Pertinent history is that the foal has had two other milder episodes with similar clinical signs within the previous two weeks but seemingly recovered without treatment each time. A spinal tap was performed which showed a mild increase in protein and macrophages with a slight

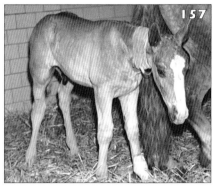

xanthachromia. All routine laboratory findings, CBC and blood chemistries, including liver enzymes, were normal except for a neutrophilia and lymphopenia (stress), and marked elevation in plasma bile acids (164µmol/l).
i. What is the most likely diagnosis?
ii. What additional blood test would you perform and what additional diagnostic procedures are needed to confirm the diagnosis?

155–157: Answers

155 i. Myotonia.
ii. Affected foals usually have progression of clinical signs, cannot be used as performance animals and are frequently given euthanasia by two years of age.
iii. EMG testing should reveal hyperactive muscle with myotonic bursts of activity.

156 The treatment is unilateral ovariectomy. The ovary is frequently cystic with large cysts containing bloody fluid (**156a**). In this mare, part of the left horn was adhesed to the ovary and this was removed at surgery (**156b**). The other ovary usually returns to normal cyclicity within 3–12 months of removing the neoplasm. Fertility should be unaffected by the presence of only one ovary. The mare in this case was artificially inseminated the following spring and became pregnant. However, she aborted in month 10 of pregnancy. Unfortunately, the fetus was not examined. It is interesting to speculate that the abortion may have been due to placental insufficiency after removal of part of the left uterine horn.

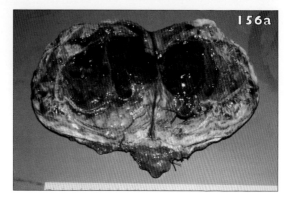

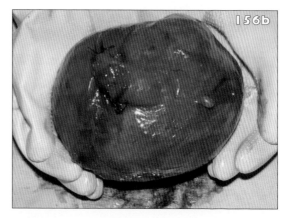

157 i. The most likely diagnosis is a portosystemic shunt.
ii. This was confirmed by measuring blood ammonia (336µmol/l) and performing a contrast portogram, which confirmed the shunt. The shunt could also be seen on a transcutaneous ultrasound examination of the foal.

158–160: Questions

158 What are the possible causes of ventricular premature depolarization (VPD)?

159 You are asked to examine a 10-week-old foal with severe, non-weight bearing lameness of the left forelimb associated with a swollen, hot and painful elbow joint (**159**). The foal is dull and inappetent, and is mildly pyrexic (38.6°C). Radiography of the joint shows no abnormalities other than soft tissue swelling.
i. What is the most likely diagnosis?
ii. How would you confirm this?
iii. What organisms are most commonly involved?
iv. What is the pathogenesis of this disease?
v. How would you treat this foal?

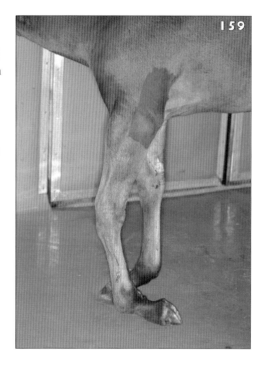

160 Shown (**160**) is an aborted foal with its placenta, which is the result of a common problem that causes abortions. What is the lesion, and what is the 'magic number' associated with the problem?

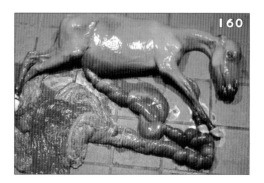

158–160: Answers

158 Infrequent VPDs can be seen in normal horses but usually occur less than five times in 24 hours. VPDs can be associated with both primary and secondary myocardial disease. Causes of primary myocardial disease include viral and bacterial infections, toxins such as monensin and drugs such as halothane, and immune-mediated or idiopathic cardiomyopathy. Secondary myocardial disease is most often seen in horses with hypoxia or gastrointestinal disorders, renal disease or septicaemia, where endotoxaemia, or electrolyte or metabolic disturbances, can induce arrhythmias. Ventricular arrhythmias also occur in association with other cardiac lesions such as endocarditis, pericarditis and aorto-cardiac fistula.

159 i. Septic arthritis ('joint ill').
ii. Synovial fluid from the affected joint is aspirated for cell count and cytology, protein estimation, Gram stain and culture. Synovial fluid from infected joints is usually turbid, and contains large numbers of inflammatory cells (primarily neutrophils), a high protein concentration and, often, floccules of proteinaceous material and fibrin. Peripheral leucocytosis and hyperfibrinogenaemia may be present.
iii. The most common isolates include *Streptococcus* spp., *Actinobacillus equuli*, *Salmonella* spp., *Escherichia coli*, *Staphylococcus aureus* and *Rhodococcus equi*.
iv. Joint infections in foals occur most commonly as a result of haematogenous spread from the umbilicus or alimentary tract. Abscesses and infections of the umbilical cord remnants are a common reservoir and source of infection. Partial or complete failure of passive transfer of colostral immunity predisposes to septic arthritis. Joint infections can also arise from penetrating joint wounds.
v. Antimicrobials should be administered systemically, the choice of drug being determined if possible by the results of bacterial culture and sensitivity testing. All presently known antimicrobials are capable of crossing the synovial membrane when administered in therapeutic concentrations. The intra-articular administration of antimicrobials may also be considered, although this might result in a chemical synovitis. Joint lavage may be necessary to drain the joint of purulent exudate and inflammatory mediators. The other joints and umbilicus should be carefully assessed for the presence of concurrent infection, and treated accordingly.

160 The problem in this case is the markedly oedematous umbilical cord, with pockets of oedema and urine within the 32 twists of the cord. The 'magic number' referred to is up to 14 twists, which is the number that can be seen in normal pregnancies. More than 14 twists associated with oedematous swelling, fluid cysts, fetal urine retention, tears and fibrosis of the umbilical cord and membranes is highly suggestive of abortion due to fetal anoxia caused by reduced placental circulation. Some cases of longer duration may show an emaciated fetus and an enlarged fetal urinary bladder and urachus.

161–163: Questions

161 Shown (161) is a sonogram of the normal right kidney of a three-month-old foal. The sonogram was performed with a 5MHz sector transducer. What are the areas identified by the dark arrow and the open arrows?

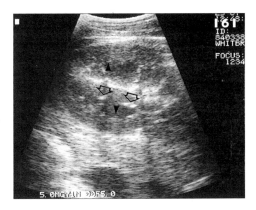

162 During the course of a routine colic examination, a large volume of gastric reflux is obtained on passage of a nasogastric tube (162a).
i. List the differential diagnoses that should be considered in relation to this single finding.
ii. Why is gastric distension a potentially life-threatening situation and how should this horse be managed if subsequently referred for further investigation and/or surgery?

163 This horse has had this hairless area for over six months (163). It has remained static in size for the previous four months. When first noticed the area was much smaller and it enlarged to its current size despite the application of a variety of antifungal agents. The skin is visually and palpably normal and no other horse on the farm has skin disease.
i. What is the most likely diagnosis?
ii. How is that diagnosis confirmed?
iii. What is the prognosis for recovery?

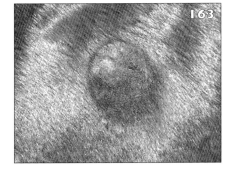

161–163: Answers

161 The open arrows indicates the corticomedullary junction. The cortex is more echogenic than the medulla. The dark arrow indicates the renal pyramids.

162 i. The causes of gastric distension leading to nasogastric reflux can be considered in three major categories:
- Ileus, eg. grass sickness, peritonitis, anterior enteritis.
- Small intestinal obstruction, eg. strangulating obstructions, non-strangulating obstructions, extra-mural obstructions (eg. nephrosplenic entrapment). The more proximal the obstruction, the earlier in the course of the disease will distension occur.
- Gastric dysfunction and/or obstruction, eg. pyrrolizidine alkaloid toxicity.

ii. Gastric distension is potentially life-threatening because, under most circumstances, the horse can not 'vomit' as other species can. This is probably due to the anatomical arrangement of the oesophago-gastric junction which acts as a non-return valve. As intra-gastric pressure increases so the sphincter becomes more tightly closed. The proximal alimentary tract of the average horse produces somewhere in the region of 120–150 litres of secretions per 24 hours: average maximum gastric capacity is approximately 20 litres. Therefore, if the gastric outflow is obstructed, it is not many hours before the stomach is tightly distended (thereby leading to ileus) and on the point of rupturing, thus releasing food material into the peritoneum and causing a rapidly fatal peritonitis. The rapidity with which the stomach refills with fluid makes it imperative that horses with gastric obstruction are decompressed immediately prior to transporting them to a referral centre. If the journey is likely to take more than two hours, it is advisable to transport the horse with a nasogastric tube in place (**162b**).

163 i. Alopecia areata.
ii. Skin biopsy. Since the diagnostic histological features of alopecia areata disappear with advancing time, biopsies should be taken at the periphery of the lesion and not in the centre.
iii. Excellent. Animals with singular lesions of alopecia areata usually regrow their hair spontaneously but regrowth can take 12 months or longer. The regrown hair is often white and this depigmentation can be transient or permanent.

164 & 165: Questions

164 You are asked to examine an 18-year-old Warmblood gelding that has developed progressive alopecia (164), which started in the neck region. Haematological evaluation reveals a haematocrit of 30% (n = 36–42%) and a normal leucocyte count (8.9×10^9/l with 21% lymphocytes, 1% eosinophils and 78% neutrophils). The total serum protein concentration is normal (81g/l with 31.7% albumin, 14.3% alpha-globulins, 29.4% beta-globulins and 24.6% gamma-globulins). Basal T_4 concentration is low.

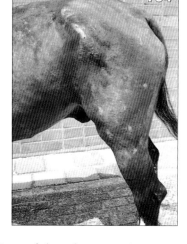

i. What is the most likely endocrinopathy to cause these lesions?
ii. Which three thyroid hormones can you name, and how are they controlled?
iii. Does diet influence the plasma concentrations of these hormones?
iv. What is the euthyroid sick syndrome?

165 A nine-year-old Quarterhorse mare is presented to you to be evaluated for acute onset of abdominal pain. Physical examination findings include temperature of 38.8°C, heart rate of 92bpm and respiratory rate of 32bpm; borborygmal sounds are decreased. During palpation *per rectum*, a few loops of distended small intestine are detected. Analysis of peritoneal fluid obtained by abdominocentesis reveals a total protein concentration of 45g/l and a nucleated cell count of 11,600 cells/µl. Passage of a nasogastric tube enables you to siphon off about 30.5 litres of fluid (165). The fluid is initially greenish in colour but the final 7.6 litres appear orange to red. After gastric decompression the horse appears to be in less pain, but within 90 minutes it exhibits signs of mild pain and the heart rate increases again. Another 9.5 litres of fluid is obtained from siphoning the stomach. This pattern of copious gastric reflux, intermittent pain and depression persists for the next several days.

i. What is the most likely diagnosis?
ii. What causes this disorder?
iii. What is the most frequent complication?

164 & 165: Answers

164 i. Hypothyroidism. Thyroid dysfunction is rare in the horse. Equine hypothyroidism can be classified into four groups: hypothyroidism in foals; hypothyroidism in adult horses; euthyroid sick syndrome; and thyroid neoplasia. Hypothyroidism in foals is usually associated with prognathism, ruptured tendons, forelimb contracture, delayed ossification of the carpal and tarsal bones, and goitre. The prognosis is extremely poor. The clinical signs of hypothyroidism in adult horses are cessation of growth, hypothermia, lethargy, irregular areas of alopecia and thickened face.
ii. Thyroglobulin is the prohormone of the thyroid hormones thyroxine (T_4), triiodothyronine (T_3) and reverse triiodothyronine (rT_3). Once released into the circulation, the hormones are rapidly bound to plasma proteins and only 0.2% T_3 and 0.06% T_4 is free or unbound. Only the free forms are biologically active. Thyroid-releasing hormone (TRH) is released from the hypothalamus in response to decreased free T_4, which in turn stimulates the anterior pituitary to release thyroid-stimulating hormone (TSH). There is a slight diurnal variation in thyroid hormone concentrations with low concentrations during the night. Thyroid hormones regulate cell growth and differentiation, and regulate energy metabolism.
iii. In weanling foals fed a diet in accordance to their energy and protein demands, both serum T_4 and T_3 concentrations increase within two hours of feeding. Undernutrition in foals results in an increase in T_4 concentration, but T_3 concentration remains unchanged. On the other hand, excess energy and protein in foals result in a decreased T_4 concentration and an increased T_3 concentration. In adult horses, food restriction induces a decrease in thyroid hormones. In addition, non-thyroidal factors affect thyroid function in horses. Three of these factors, age, pregnancy and low temperature, are associated with increased levels of thyroid hormones. Thyroid hormone concentrations in normal foals are up to 14 times higher compared with those of horses older than two years. Serum thyroid hormone concentrations are decreased following concurrent disease and the administration of phenylbutazone (T_4 only). However, the administration of dexamethasone does not change baseline T_3 and T_4 concentrations.
iv. The euthyroid sick syndrome occurs in horses. Disease unrelated to thyroid disease can depress basal thyroid hormone levels as a normal response to minimize the catabolic effects of thyroid hormone during disease. In cases of euthyroid sick syndrome, low baseline T_3 and T_4 concentrations respond normally to TSH administration.

165 i. Duodenitis/proximal jejunitis (also known as anterior enteritis).
ii. The cause remains unknown, although infection with clostridial organisms and *Salmonella* spp. and mycotoxicosis have been suggested.
iii. The most common complication is laminitis, developing in approximately 25–30% of cases.

166–168: Questions

166 A 16-year-old Irish Draught-cross mare is presented for pregnancy diagnosis in July. This year she had been covered at three oestrus cycles and has been diagnosed pregnant at an ultrasound scan carried out in May, 21 days after her previous cover. At that time she was sent home. Now (July) she is scanned by you and found to be non-pregnant.

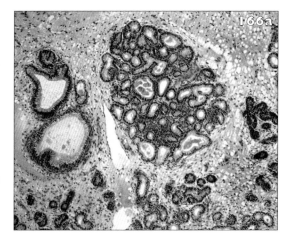

i. What further details about her reproductive tract would you check by ultrasonography and by manual palpation per rectum?
ii. An endometrial biopsy is collected (**166a**) – what pathological change(s) are present?
iii. Could these have caused early fetal death?

167 You are performing a pre-purchase (soundness) examination on a five-year-old Warmblood gelding. The left eye appears as in **167**. What is your diagnosis?

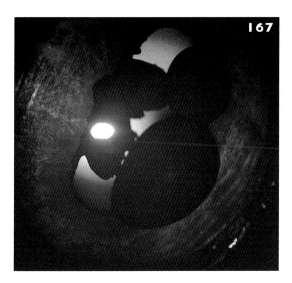

168 i. What is the cause of glanders?
ii. Geographically, where does glanders occur?

166–168: Answers

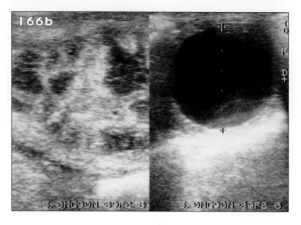

166 i. Palpation will reveal the normality of morphology of the uterus and ovaries. The size and shape of the ovaries should be noted along with the presence of any follicles. The uterus should be carefully palpated for shape and tone. Lymphatic cysts are common in older mares – possibly through decreased myometrial tone. Mural cysts are commonly located ventrally at the uterine horn–body junction and can be palpated as a ventral sacculation. If the cysts are sufficiently widespread to give a doughy feel to the uterus, it is thought that fertility is decreased.

Ultrasonographic examination of the uterus should detect whether fluid is present in the uterus. Fluid in the uterus during dioestrus, or excessive amounts of fluid during oestrus, are highly suggestive of an endometritis. The stage of the oestrus cycle can be determined by noting the size of follicles on the ovaries, detecting the presence of a corpus luteum and recording the presence or absence of uterine oedema. The tone of the cervix may also help in determining stage of cycle. In oestrus, a follicle or follicles of greater than 30mm in diameter will be present in the ovaries (**166b**, right); uterine oedema will be detected (**166b**, left) and palpation of the cervix per rectum will reveal softening. However, in maiden mares the cervix may not relax. In dioestrus, a corpus luteum may be detected on scan and follicles of varying sizes may be present. The uterus will appear relatively homogeneous with no oedema. The cervix will be long, narrow and hard due to the effect of progesterone.
ii. The changes seen are mild periglandular fibrosis with associated cystic glandular distension. These changes occur with increasing frequency in older mares and are classed as chronic and irreversible. Widespread, severe fibrosis is associated with very low chances of the mare carrying a foal to term.
iii. Periglandular fibrosis is thought to be a common cause of embryonic and early fetal death between days 40 and 90.

167 Iris cysts. Although most horses with iris cysts are asymptomatic, some horses have cysts large enough to obscure a significant portion of the pupil, thereby interfering with vision. Horses with one cyst frequently develop others in the same or opposite eye.

168 i. *Pseudomonas mallei*.
ii. The disease is confined to a few regions of Asia.

169 & 170: Questions

169 An eight-year-old Standardbred mare (**169a**) is presented with a one-week history of mild colic, depression, partial anorexia and intermittent fever (as high as 39.7°C). She is the only horse on this large farm demonstrating these clinical signs. She has been treated with trimethoprim/sulfa (20mg/kg p/o bid) since the onset of clinical signs seven days previously. The mare has also received flunixin meglumine as needed for the colic. Examination reveals the following: T, 39.7°C, HR, 45bpm, RR, 16bpm. The sclera are yellow (**169b**) and the mucous membranes of the mouth appear orange, capillary refill time is abnormally slow and the mucus of the mouth is very tenacious. The mare also appears thin. Pertinent laboratory abnormalities include: PCV, 49%; plasma protein, 92g/l; WBC, 16.2x10^9/l; neutrophils, 13.7x10^9/l; lymphocytes, 2.1x10^9/l; monocytes, 0.4x10^9/l; plasma fibrinogen, 7.0g/l; urea nitrogen, 12.2mmol/l (34.2mg/dl); GGT, 474u/l; AST, 660u/l; total bilirubin, 225μmol/l (13.2mg/dl); conjugated bilirubin, 71.8μmol/l (4.2mg/dl); serum globulin, 58g/l. The urine is dark orange, strongly positive for bilirubin, has trace occult blood and a specific gravity of 1.032. The sonogram of the liver is shown (**169c**).

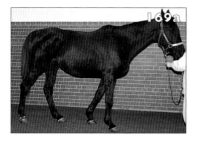
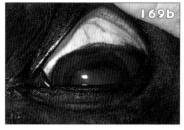
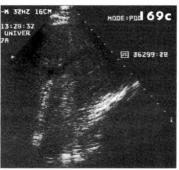

i. What is the most likely diagnosis in this case?
ii. Would a liver biopsy be indicated?

170 You are asked to examine a four-year-old trotter mare with a history of bilateral haemorrhagic nasal discharge after each race.
i. What is your tentative diagnosis?
ii. How would you confirm the diagnosis?
iii. How and where does the damage develop in the lungs?
iv. Will this disease influence the exercise tolerance or performance ability of this horse?
v. How would you treat this animal?

169 & 170: Answers

169 i. The most likely diagnosis is cholangiohepatitis with sludging of bile and/or concretions (acoustic shadows) in the biliary tract. The marked elevation in GGT, conjugated bilirubin, fibrinogen and urea nitrogen suggest that the disease is predominantly in the biliary tract.
ii. A liver biopsy should be performed to obtain a piece of liver for culture. The diagnosis is not in doubt but the offending organism needs to be identified. In this case, an *Enterobacteria* sp. was cultured that was resistant to trimethoprim/sulfa. The mare was treated with enrofloxacin and recovered.

170 i. Exercise-induced pulmonary haemorrhage (EIPH).
ii. Endoscopic examination carried out 30–60 minutes after exercise will reveal blood in the trachea and large bronchi (**170**). Cytological examination of tracheal aspirates or bronchoalveolar lavage samples will reveal haemosiderophages (macrophages containing green pigment; haemosiderin may be stained specifically using Pearl's Prussian Blue stain). These cells can be identified in samples obtained up to three months after pulmonary haemorrhage. In the acute stage of extensive EIPH, a lateral radiograph of the dorsocaudal lung area may reveal an increased bronchointerstitial pattern or a localized radiopacity. In many cases, however, radiographic findings are disappointing.
iii. The bleeding originates from the pulmonary capillaries. During high speed exercise, the pressure difference between the pulmonary capillaries and the alveolar space can result in mechanical failure of the capillary wall.
iv. Data from large surveys indicate that EIPH occurs in more than 70% of all racehorses. There is no obvious difference in the incidence of the condition between successful and unsuccessful animals. Although EIPH will influence the performance ability of some individual horses, many authors consider it to be an inevitable consequence of the high cardiac output in elite athletes.
v. In many countries, frusemide (furosemide) is used in an attempt to prevent EIPH, or at least to decrease the amount of blood loss. However, a number of surveys have failed to show a positive effect of frusemide on the occurrence of EIPH or on the ability of treated animals. Since EIPH seems to be an inevitable consequence of high level performance, the best way to prevent extensive blood loss is to minimize damage to the airways caused by respiratory viruses or environmental pollutants (eg. stable dust, ammonia, etc.). Many clinicians advocate a period of rest for horses that have been affected by extensive blood loss due to EIPH. Unfortunately, this interferes with training schedules and the owner's interest in the horse. Other clinicians suggest that affected horses should be kept in light work following an episode of haemorrhage.

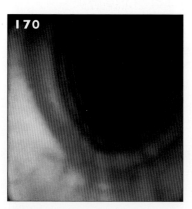

171–173: Questions

171 i. How is equine infectious anaemia transmitted?
ii. What are the clinical signs of the acute and chronic forms of equine infectious anaemia?

172 A three-year-old Standard-bred filly acutely developed severe generalized muscle tremors and a short strided gait without ataxia, then collapsed in her stall at the race track. On clinical examination the filly was bright and alert but had easily detectable decreased tone in the tongue (**172**), eyelids, tail and anus. She could stand but would remain standing for only 10–20

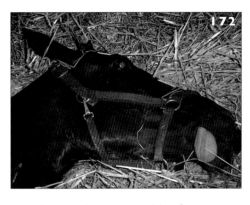

minutes. Her appetite was excellent but food and water would reflux out the nostrils, so intake was withheld. Five days earlier, a stable mate had died after exhibiting similar clinical signs for 36 hours.
i. What is the diagnosis?
ii. What is the appropriate treatment?
iii. How does the epidemiology of the disease differ between horses and foals?

173 A four-year-old pony from the UK was presented for investigation of diarrhoea. The diarrhoea initially resolved following treatment with oral codeine phosphate but recurred several days later. Ten days after she was first ill the pony is in poor body condition, dull, inappetent and diarrhoeic (**173**), but there are no specific abnormalities on physical examination. Hypoalbuminaemia (16g/l), hyperglobulinaemia (42g/l), elevated plasma alkaline phosphatase (546u/l) and neutrophilia (17.9×10^9/l) have been identified. No significant bacterial pathogens have been cultured from two separate faecal samples submitted on the third and fourth days of illness and no parasite eggs have been identified in these samples. What is the most probable diagnosis?

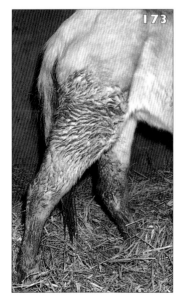

171–173: Answers

171 i. The virus of equine infectious anaemia is transmitted by biting flies, or by contaminated needles, teeth rasps, stomach tubes and other instruments. Foals may be infected *in utero* by transplacental infection, or postnatally by ingesting virus-contaminated colostrum or milk.
ii. The acute form is characterized by fever, oedema of dependent areas, haemorrhagic diarrhoea, jaundice, petechial haemorrhages, anaemia and death. The chronic form is characterized by progressive anaemia, weight loss, weakness and oedema; recurrent episodes of clinical disease are common.

172 i. Botulism.
ii. The filly should receive treatment with *Clostridium botulinum* antiserum. Antibiotics should be given since most cases have some aspiration pneumonia. Aminoglycosides and procaine penicillin should be avoided since they may further diminish neuromuscular transmission. Nutritional support should be provided by tube-feeding a home-made gruel (alfalfa meal, electrolytes, dextrose, corn oil) or a commercially prepared enteral diet. This should be fed until the dysphagia is resolved. Excellent nursing care is needed to prevent decubital sores, etc.
iii. In adult horses the disease is usually acquired from ingestion of the preformed toxin in the feed. In young foals (generally 2–8 weeks of age) the disease is a result of overgrowth of the organism in the lower bowel and *in vivo* production of toxin (enterotoxigenic). *Cl. botulinum* is usually present in the faeces of affected foals and a positive culture is generally considered diagnostic in foals.

173 Larval cyathostomosis. This condition is a fairly common clinical entity in Europe but apparently less common in other continents. Typically, this is a winter–spring onset condition of animals between one and five years of age but cases are seen at other times and in older age groups. The condition constitutes a protein losing enteropathy as a result of typhlitis and/or colitis associated with large numbers of mucosal stages of cyathostome parasites. Not all cases develop diarrhoea and some may show additional features of low-grade colic and/or intermittent fever and/or peripheral oedema. There is evidence that recent (two weeks) anthelmintic dosing induces resumption of development of larvae which had been and inhibited which may evoke clinical disease.

The clinicopathological findings in this case are typical of cases of larval cyathostomosis but they are not pathognomonic of the condition. The condition is caused by larval stages of cyathostomes which do not produce eggs in faeces.

174–177: Questions

174 This ultrasound image (**174**) was obtained from a Standardbred yearling with a murmur.
i. What is the diagnosis?
ii. Describe the typical characteristics and locations of any murmur(s) that could be associated with this problem.
iii. What are the range of sequelae associated with this problem?

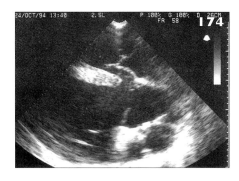

175 i. What disease is associated with infection by *Borrelia burgdorferi*?
ii. How is this organism transmitted?
iii. What are the major clinical signs associated with this disease?

176 A five-month-old foal is found down in the pasture with acute unilateral blindness. Due to the poor prognosis, euthanasia was performed. On post-mortem examination, the brain was found to have a large right cerebral lesion (**176**).
i. In which eye would the blindness occur?
ii. What is the most likely aetiologic agent?

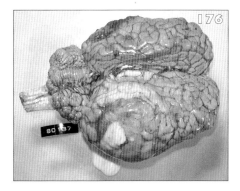

177 A yearling Shire filly is examined because of urinary incontinence since birth. There is no dysuria and the filly will void urine normally, but there is also an almost constant dripping of urine from the vagina. A cystoscopic examination is performed. The serum creatinine of the filly was normal, 97.2µmol/l (1.1mg/dl).
i. What are the findings shown (**177**) and what is the diagnosis?
ii. What are the treatment options for correcting this problem?

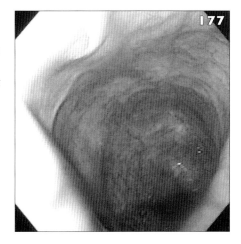

174–177: Answers

174 i. Ventricular septal defect (VSD).
ii. There are typically systolic murmurs over both the left and right sides. The murmur on the left is loudest over the pulmonic valve area (far forward, well under the triceps over the third intercostal space, and just above the point of the elbow). The murmur on this side is usually ejection to band-shaped in quality, and grade III to IV/V. The murmur on the right is often louder than that on the left, is band-shaped in quality, and loudest over the tricuspid valve area (4th intercostal space on the right, between the elbow and point of the shoulder).
iii. Because the VSD is relatively small (<2.5cm diameter), it is unlikely to ever be haemodynamically life-threatening (ie. this is a restrictive VSD). However, a small VSD may none the less reduce a horse's inherent athletic ability. Certainly, race horses have competed with ventricular septal defects; however, their performance may have been less than it would have been otherwise. VSDs that are 'non-restrictive', ie. that are of haemodynamic importance, will cause volume overload of the left side of the heart. (However, if a large VSD is located in the muscular portion of the septum, the right heart will become dilated.) Subsequent to this, atrial fibrillation and/or secondary mitral regurgitation due to dilation of the mitral valve annulus can occur. Ultimately, congestive heart failure will result from large VSDs. Some defects may develop fibrosis around the orifice over time, in effect reducing the size of the defect.

Occasionally, there may be aortic regurgitation of varying severity if the VSD results in loss of support of the right coronary cusp of the aortic valve. The aortic valve cusp may prolapse into the defect during diastole, resulting in aortic regurgitation.

If pulmonary hypertension occurs, either independently or as a result of the VSD, increased right ventricular pressure will develop. Either decreased left to right flow through the VSD or even reversed (right to left) flow can, therefore, ensue. Although VSDs alone are the most common congenital cardiac defect in the horse, VSDs often accompany other types of congenital cardiac defects. A thorough echocardiographic examination should be performed with this in mind.

175 i. Lyme disease (borreliosis).
ii. *B. burgdorferi* is transmitted by nymph and adult ticks of the *Ixodes* group. The organism may also be transmitted by blood, urine and synovial fluid.
iii. Lameness, lymphadenopathy, fever, inappetence, uveitis, neurological signs.

176 i. Left eye visual deficits.
ii. *Streptococcus equi* subsp. *equi*.

177 i. Shown (**177**) is the air-distended bladder with a small amount of urine in the ventral aspect and the opening of the right ureter (reverse image). There is no ureteral opening into the bladder for the left ureter. The diagnosis is ectopic ureter (left).
ii. The two options for correcting the problem would be either nephrectomy of the left kidney or transposition of the ectopic left ureter into the bladder. Transposition was successfully performed in this case.

178–180: Questions

178 The foal pictured in **178** has contracted tendons and requires assistance to get up. What medical treatments, physical treatments or surgical therapy can be performed?

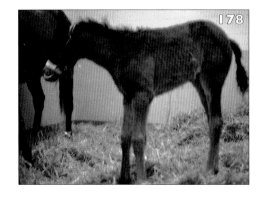

179 A 14-year-old Thoroughbred gelding has a six-year history of a chronic cough and an intermittent nasal discharge. Thoracic auscultation revealed loud end expiratory wheezes. A tracheobronchial aspirate specimen is seen in **179**.
i. What are the cell types seen?
ii. What is the likely diagnosis?
iii. What treatment is required?

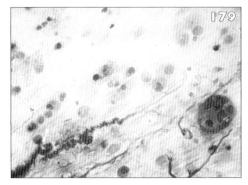

180 The two large pieces of large colon shown in **180** are from two different horses with diarrhoea and fever. The lighter-coloured necrotic mucosa may be seen in horses anywhere in the world and at any time of year, while the darker-coloured oedematous bowel is commonly seen in certain specific areas of

the USA, usually during the warmer season of the year. There is often a definite 'septic tank' odour to the pale mucosa, but no specific odour to the darker bowel. Both horses had died as a result of their bowel lesions.
i. Name the diseases most likely to be involved in each instance.
ii. What two other diseases must be differentiated with the darker mucosa?

178–180: Answers

178 Muscle relaxants are usually not effective, and the use of oxytetracycline offers more consistent results. The mechanism by which the tetracyclines cause tendon relaxation are not completely understood. Physical therapy can include periodic stretching of the limbs while sedated or when the foal is recumbent. Casting and splinting may provide further support for ambulation. Surgical procedures include flexor tenotomy in severe cases and check ligament desmotomy in less severe cases. Foals which fail to extend the digit, although a response to contracted tendon treatments appears to be beneficial, may have secondary rupture of the extensor process of the long digital extensor tendon at the attachment process of the pedal bone.

179 i. Macrophages, a Curshman spiral, a multinucleated giant cell, respiratory epithelial cells, mucus and debris.
ii. Chronic obstructive pulmonary disease (COPD).
iii. The key means of controlling COPD is by reducing the horse's exposure to the offending agents, most often airborne dust and moulds. Pharmaceutical intervention may be necessary and could include the use of bronchodilators, mucolytics and anti-inflammatory agents. Bronchodilators may include drugs from these three groups: muscarinic receptor antagonists – parasympatholytics; β_2-adrenoceptor agonists – sympathomimetics; and methylxanthine derivates. Mucokinetic drugs increase the transport of the complex and incompletely understood airway secretions by differing mechanisms. Mucokinetic agents to consider in the horse include bromhexine, dembrexine, theophylline and clenbuterol. Anti-inflammatory drugs primarily affect the delayed response and bronchial hyperreactivity. Corticosteriods most commonly used are prednisolone and dexamethasone.

180 i. The lighter-surfaced colon with the 'septic tank' odour suggests a diagnosis of salmonellosis, and these findings are incriminating in themselves, even if *Salmonella* spp. are not cultured. The darker, seriously oedematous colon is congested, and the time of year is suggestive of an arthropod-carried disease such as Potomac horse fever (equine monocytic ehrlichiosis). The diagnosis should not be based solely on these findings.
ii. Other diseases which may cause similar lesions in the large colon to the darker bowel, but which may occur at any time of year, include vascular compromise due to torsion, clostridiosis, displacement or strangulation. Uraemic colitis has a similar appearance, but has a characteristic uraemic smell due to renal failure. Colitis X should also be considered if the history is appropriate (such as extreme environmental exposure, long distance travel, etc.).

181 & 182: Questions

181 The 16-year-old mare described in **166** has aborted between 21 and 60 days after conception, but the owner is still keen to breed the mare this year.
i. What criterion will determine how soon the mare returns to oestrus? If she comes back into oestrus what further tests might you want to perform prior to service?

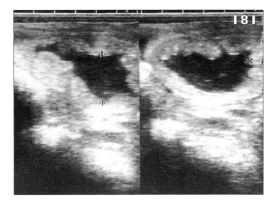

ii. The owner detects the mare in oestrus. She is rescanned with the result shown in **180**. What is the likeliest diagnosis?
iii. How would you treat and manage this mare?

182 The ultrasound (**182a**) of the umbilical remnant at the level of the urachus was obtained prior to excision from a three-week-old foal with a patent urachus. The excised remnants at the level of the bladder apex are shown in **182b**. What is the diagnosis?

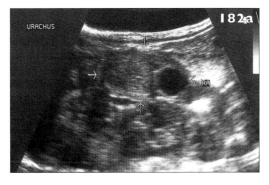

143

181 & 182: Answers

181 i. Endometrial cups form in the uterus by invasion of the endometrium by cells from the chorionic girdle of the embryo at around days 35–40 of pregnancy. The endometrial cups secrete equine chorionic gonadotrophin (eCG; formerly known as PMSG). eCG is thought to stimulate steroidogenesis in the maternal ovaries. If the embryo is lost after day 35–40, the endometrial cups have already formed and maternal ovarian progesterone output is maintained as though the mare were pregnant. Thus, if the conceptus is lost after day 35–40, it is unlikely that the mare will come back into oestrus until the endometrial cups have regressed between days 120 and 140. By this time it is likely that the mare will have entered seasonal anoestrus.

Further investigation of this mare prior to breeding should comprise a thorough evaluation of the vulval and perineal conformation, collection of an endometrial swab and cytological examination of uterine contents. As swabs can be subject to contamination, evaluation of uterine cytology is extremely useful in determining whether inflammation is present.

ii. The scan picture shows fluid in the uterus. Even though the mare is in oestrus, this is a greater amount of fluid than would be deemed normal. It is highly likely that the mare has endometritis.

iii. Either oxytocin (25iu) should be administered intravenously to empty the uterus, followed 60 minutes later by infusion of water-soluble antibiotic in a volume of 20–100ml depending on the size of the mare's uterus, or, preferably, the uterus should be flushed out with one litre volumes of sterile saline until the recovered fluid is clear. Oxytocin and antibiotic therapy should be applied as above. It is important that this treatment is continued daily for 3–5 days until no fluid is present on scan. Endometritis should be treated only during oestrus when uterine contractility and immune defence mechanisms are optimal.

It is easier to control endometritis in mares when artificial insemination is used rather than natural service (less bacterial contamination). Mares should always be scanned on the day after insemination to check for fluid accumulation. In mares susceptible to endometritis it is usual to treat the endometritis resulting from insemination as described above, ie. either oxytocin + antibiotics or flush + oxytocin + antibiotics. The embryo takes five days to reach the uterus and therefore treatment can be repeated daily from insemination until day four after ovulation. Treatment should be stopped when no fluid is detected in the uterus on ultrasound scan.

Note. It is important that only mares which have evidence of recurrent endometritis are treated in this way. Pregnancy rates in most mares are very good without treatment when they are bred at the correct time using a fertile stallion. It is normal for every mare to develop a certain degree of endometritis after breeding.

182 Clot in the right umbilical artery. The clot was sterile and an impression smear revealed no inflammatory cells or bacteria. This example illustrates the importance of distinguishing between a benign sterile lesion and infection of internal remnants. The sonographic interpretation must be made with consideration of the patient history and sonographic changes in the other umbilical remnants.

183–185: Questions

183 This seven-year-old gelding (183) has been passing blood at the end of urination for two weeks. A rectal examination is normal, there is no dysuria and a catheterized urine sample is normal.
i. From where is the blood originating?
ii. What diagnostic procedures would you perform?

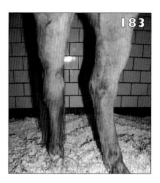

184 A yearling Shetland pony mare is examined because it has been sluggish and anorexic for two days. Some *Digitalis purpurea* plants are present on the pasture. The tentative diagnosis of digoxin intoxication is made on the basis of heart auscultation and an ECG, and confirmed by a plasma digoxin concentration of 1.4μg/l. The serum aspartate aminotransferase activity is elevated (510u/l; n = 250–350u/l), as are the activities of serum gamma–glutamylaminotransferase (99u/l; n = 10–40u/l) and lactate dehydrogenase (1,831u/l; n = 350–650u/l).
i. What complication is most likely in view of the serum enzyme activities?
ii. How would you treat this complication?
iii. How could this treatment exaggerate the cardiac arrhythmia?
iv. What rapidly fatal complication can result from this metabolic disturbance?

185 A two-year-old Standardbred from the US is examined because of poor training and loss of muscle mass over the right gluteal area (185a). On clinical examination additional findings were scuffing of the toe of the right hindlimb and atrophy of the tongue on the right side (185b). Fasciculations were noted in the tongue. There was no apparent ataxia.

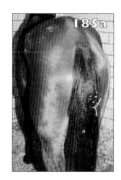
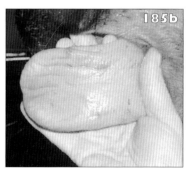

i. What is the most likely diagnosis?
ii. What ante-mortem test would be recommended to confirm the diagnosis?
iii. What would the proper treatment be for this condition?

183–185: Answers

183 i. Bleeding at the end of urination in the male horse is highly suggestive of a urethral disorder. The bleeding occurs at the end of urination as the horse has muscular contractions of the urethra.
ii. Endoscopic examination of the urethra should be performed. Cultures of the urethral mucosa are usually not helpful. Generally, two types of lesions are seen on urethroscopic examination (using a 1m endoscope): prominence of urethral vessels and hyperaemia in the dorsal urethral area at the openings of the accesory sex glands or diffuse redness of the urethra with linear erosions in some areas. Healing occurs in most cases within several weeks. Antibiotics are generally administered but their value is unclear, especially in those cases with enlarged tortuous vessels in the proximal urethra.

184 i. The serum enzyme activities indicate hepatic damage, probably caused by hyperlipaemia. Hyperlipaemia has been reported most frequently in Shetland and Miniature ponies, and in donkeys. Lipid metabolism in the pony seems to be different from that in the horse. Ponies and burros will develop visibly lipaemic plasma after only four days of fasting, whereas horse plasma becomes only slightly turbid towards the 18th day of fasting.
ii. Specific treatment is aimed at decreasing the negative energy balance by nutritional support via nasogastric tube at least twice daily. Heparin (up to 200iu/kg bodyweight s/c or i/v sid) can be administered to increase the rate of lipid clearance by enhancing the plasma activity of lipoprotein lipase. Lipoprotein lipase is bound to endothelial cells by heparan sulphate proteoglycan. By competing with the heparan sulphate bond, heparin releases lipoprotein lipase, thereby increasing its activity. However, it has been suggested that the activity of lipoprotein lipase is already near maximum in ponies with hyperlipaemia. Protamine zinc insulin (up to 0.4iu/kg s/c sid) and ultralente insulin (0.4iu/kg i/v sid) in combination with i/v administration of 10% glucose can be used to inhibit fat mobilization by inhibiting hormone-sensitive lipase, and to increase triglyceride uptake into peripheral tissues by stimulating lipoprotein lipase. The overall survival rate in equine hyperlipaemia is around 50%.
iii. The administration of insulin facilitates the influx of K^+ into cells, which in turn can lead to hypokalaemia, thereby enhancing the cardiac arrhythmia.
iv. An intra-abdominal haemorrhage due to liver rupture as a result of hepatic lipidosis.

185 i. Equine protozoal myeloencephalitis (*Sarcosystis neurona*).
ii. A western blot antibody test or polymerase chain reaction to detect the antibody to or DNA of *S. neurona*.
iii. Treatment should include trimethoprim/sulfa (20mg/kg p/o bid). In addition, pyrimethemine (1.0mg/kg sid) may be given *per os*, and dimethyl sulphoxide (1g/kg sid i/v) may be given in more acute cases.

186–188: Questions

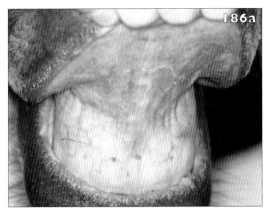
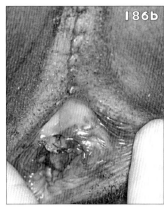

186 This is an eight-year-old Hanoverian mare in which thrombocytopenia was discovered as an incidental finding when a CBC was performed prior to anaesthetizing the mare for an elective ovariectomy. Upon physical examination, multiple petechial haemorrhages were discovered on the oral mucous membranes and vulva (**186a, 186b**).
i. What is your explanation for this finding?
ii. If extensive diagnostic testing reveals no underlying disease or possible predisposing cause, what is your recommendation to this owner, whose primary interest is the ovariectomy?

187 i. What is the cause of Potomac horse fever?
ii. What are the clinical features of this disease?

188 A mare is presented for pregnancy diagnosis 16 days after her last covering date. On lifting her tail, multiple vesicles and pustules are seen on the vulvar lips and perineal skin (**188**). What is your diagnosis, how would you treat the mare and how would you control the disease?

186–188: Answers

186 i. The thrombocytopenia is probably clinically significant, since the horse has petechial haemorrhages, even though no signs of bleeding have been noticed. Immune-mediated thrombocytopenia (IMTP) is most likely. The sensitizing agent can be infectious (bacterial or viral, such as equine infectious anaemia virus), neoplastic (lymphosarcoma) or pharmacological. Often, the condition is considered idiopathic.
ii. Surgery cannot be performed safely at this time. The IMTP is most likely idiopathic in this horse. These animals usually respond to immunosuppressive corticosteroids and, once in remission, may remain there for prolonged periods of time without medication, at which time the ovariectomy could be safely completed. In horses that do not respond to corticosteroids, azathioprine has been administered safely and successfully.

187 i. *Ehrlichia risticii*.
ii. Clinical signs may include any combination of the following: fever, depression, anorexia, ileus, colic, diarrhoea and laminitis. There is considerable variation in the clinical manifestations of the disease in individual horses.

188 These lesions are characteristic of equine coital exanthema, or equine herpesvirus type 3 infection. This is a viral disease which affects both sexes and is transmitted at coitus. The incubation period is 4–7 days; initially, multiple nodules appear on the vulva and perineum. The nodules become vesicles and pustules and eventually ulcerate leaving bare areas approximately 3–10mm in diameter. The condition resolves in 2–3 weeks leaving non-pigmented areas where skin has healed. In cases where secondary infection occurs the pustules can merge and the area becomes oedematous and covered in a mucopurulent exudate. Lesions on the stallion start as vesicles on the penis and prepuce which ulcerate. At this stage he is usually reluctant to breed mares. The condition may be preceded by transient pyrexia and depression and is only contagious during the first 10–14 days of the disease, ie. the acute phase. Infection does not confer permanent immunity, although stallions are not usually reinfected during a single breeding season.

Inspection of the lesions, along with a history of recent sexual contact, is usually sufficient for diagnosis. Confirmation can be by virus isolation or detection of herpesvirus intranuclear inclusion bodies in cytological smears or histological preparations. Serology for antibody can also be performed on paired samples collected at an interval of 2–3 weeks.

Control is achieved by preventing contact with infected animals. Animals must be sexually rested for three weeks or until the ulcers have healed. Local treatment for at least three consecutive days with an antiseptic cream is useful to help prevent secondary infection.

189 & 190: Questions

189 A five-year-old pony mare was presented with continuous signs of moderate colic of at least two hours duration (**189**). The signs were first noticed when the pony was brought in from pasture in the evening after being apparently normal in the morning. There was excess salivation, and a few millilitres of brown fluid refluxed from the

nares. Water poured from the mouth when the pony attempted to drink. Generalized sweating was present and there were fine muscle tremors over the triceps area and flanks. The pulse rate was 84/minute. The owner reported that the abdomen was becoming increasingly distended. No urine or faeces had been passed since the signs were noticed and the pony was inappetent. On rectal examination, firm faecal pellets were present in the small colon and a firm impaction of the pelvic flexure was palpable.
i. If this pony is located in eastern Scotland, what are the main differentials and what is the most likely diagnosis?
ii. How would you investigate the case further with a view to making a definitive diagnosis?

190 This healthy two-year-old horse is presented in early summer for a moth-eaten alopecia (**190**). The horse is not pruritic. The owner initially noticed a few lesions in the saddle region. Despite bathing with iodine, the lesions worsened and now cover most of the horse's trunk. Careful inspection of the skin reveals no primary lesions, only multifocal

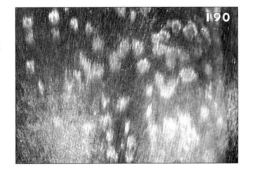

areas of hair loss with peripheral epidermal collarettes. Hairs are plucked for a fungal culture on dermatophyte test medium (DTM) and the plate turns red with a greenish-black growth in seven days.
i. The history and fungal culture results do not support a diagnosis of dermatophytosis. Briefly explain why not.
ii. Give two differential diagnoses for the condition.
iii. Which one diagnostic test would confirm your diagnosis?

189 & 190: Answers

189 i. The pony has presented with continuous colic of fairly sudden onset, although the exact duration was unknown. The following conditions should be considered in the differential diagnosis: primary or secondary gastric dilation or impaction; small intestinal obstruction due to intraluminal obstruction (eg. ileal impaction, foreign bodies); intramural lesions (eg. haematoma, oedema or neoplasia); extramural compression (eg. small intestinal volvulus, entrapment through natural or pathological openings including the epiploic foramen or tears in the mesentery); compression from a pedunculated lipoma (unlikely in a five-year-old), or small intestinal intussusception; large intestinal obstruction including primary or secondary large colon impaction, colonic displacements, caecal impaction, caecocaecal or caecocolic intussusception; functional obstruction, eg. paralytic ileus associated with peritonitis, proximal enteritis, ragwort poisoning or grass sickness (equine dysautonomia). Other causes of dysphagia could be considered, especially neurogenic causes such as botulism or pharyngeal paralysis associated with guttural pouch mycosis, although cases with dysphagia from these causes are unlikely to show colic concurrently. In this pony the presence of excess salivation, inability to swallow water, nasogastric reflux and a firm impaction of the large colon which developed while the animal was on a mainly grass diet is strongly suggestive of acute grass sickness.

ii. Further investigation should include passage of a nasogastric tube, peritoneal fluid analysis and possibly exploratory laparotomy. On nasogastric intubation, varying amounts of green to brown fluid with a characteristic putrid odour are obtained in acute grass sickness. This is useful diagnostically but can sometimes resemble the reflux obtained in other obstructive conditions. Peritoneal fluid analysis is of value in helping to decrease the likelihood that an ischaemic intestinal lesion is present. In acute grass sickness the fluid is usually deep yellow (similar to other causes of medical colic), but has a higher protein content than most other medical colics (except peritonitis). Peritoneal fluid analysis can therefore be useful, but cannot lead to a positive diagnosis of grass sickness. In this case, exploratory laparotomy should be considered if the clinician is not sufficiently certain of the clinical diagnosis. Ileal biopsy at laparotomy is currently the only means of confirming an ante-mortem diagnosis of grass sickness.

190 i. Despite the colour change in an appropriate period of time, dermatophyte colonies are not greenish-black. Clinically it would be very unlikely for a healthy horse to develop such an explosive ringworm infection during the summer.
ii. Staphylococcal folliculitis, sterile pustular disease like pemphigus foliaceus.
iii. Without any primary lesions, skin biopsy is the only reliable diagnostic test to describe this horse's skin condition.

191–193: Questions

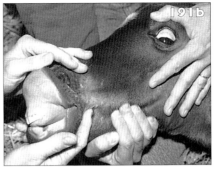

191 One of the most common clinical entities in the newborn foal is the presence of fractured ribs. **191a** demonstrates the examination procedure for palpating the ribs for oedema, asymmetry and crepitus. **191b** shows a foal with pale mucous membranes in a haemorrhagic crisis.
i. What anatomical structures can cause haemothorax and haemoperitoneum secondary to fractured ribs?
ii. What would be a cause of sudden death with or without haemothorax secondary to fractured ribs?
iii. What is a 'flail' chest?
iv. What is the treatment for fractured ribs?

192 The four-year-old pony described in **173** is suspected of having larval cyathostomosis, which has resulted in weight loss and diarrhoea. Blood analyses reveal hypoalbuminaemia and neutrophilia. How would you investigate and treat this case?

193 A 13-year-old Standardbred mare is examined because of an acute onset of abnormal behaviour, which includes propulsive circling in either direction, anorexia, wall biting and head pressing. On clinical examination the mare appears ataxic and has a slow menace response with normal pupillary light response. The mucous membranes appear discoloured (**193**).

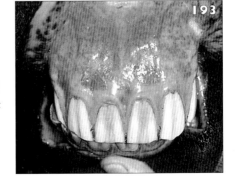

i. List three broad physiological or pathological causes of orange discolouration of membranes in the equine.
ii. Which one is least likely in this case?

191–193: Answers

191 i. Laceration of the lung, epicardium or pericardium can cause haemothorax. Haemoperitoneum is usually the result of a lacerated diaphragm which results in a diaphragmatic hernia.
ii. Cardiac laceration, either superficial or deep.
iii. The chest cavity will indent during inspiration, usually along the sites of the rib fractures. The affected lung is collapsed.
iv. Stall confinement for at least two weeks. Acute, severe cases may require sedation for non-ambulation. The foal should lay recumbent with the fractured rib side in the down position. Extensive nursing and supportive care are essential until the patient stabilizes.

192 Additional investigations which should be undertaken are:
- Microscopy of wet faecal preparation – cyathostome larvae can be identified in large numbers; indeed, these can often be appreciated as small (0.5cm) white or red worms by gross inspection of a faecal sample.
- Serum protein electrophoresis – the presence of betaglobulinaemia supports a diagnosis of cyathostomosis but this finding has limited specificity and sensitivity for cyathostome infection, ie. there may be false positives and/or false negatives.
- Intestinal biopsy – rectal biopsy is easy and safe but only positive evidence of cyathostomes in mucosal samples are helpful and a negative result does not rule out the diagnosis. Intestinal biopsy at laparotomy would be the ideal method of diagnosis but is rarely feasible on grounds of cost and risk of surgery in an hypoalbuminaemic animal.

There are three general components of treatment of larval cyathostomosis:
- Anthelmintic therapy options are as given in 223 (iii).
- Corticosteroid therapy appears useful for clinical cyathostomosis cases. The benefit may arise directly from anti-inflammatory effects but it has been hypothesized that the inhibition of larval development is immune mediated and that suppression of intestinal immune mechanisms may allow resumption of larval development, at which point the efficacy of anthelmintics is greater than during the inhibited state. Corticosteroid regimens of initial parenteral dexamethasone (0.1mg/kg bodyweight) or prednisolone (1mg/kg) followed by tapering dose of oral prednisolone over a 2–3 week period are apparently beneficial in affected cases.
- Antidiarrhoeal medication with oral codeine phosphate (up to 3mg/kg bodyweight tid) is usually effective but may have to be maintained (on a reducing dose) for several weeks. Nutritional support and/or oral probiotics or transfaunation are worth considering for cyathostomosis cases.

193 i. Hepatic failure, haemolytic disease, physiological icterus associated with anorexia.
ii. The least likely one of the three in this case is icterus caused by anorexia. Icterus caused by anorexia is rarely this severe and generally requires 2–3 days of anorexia to be noted clinically.

194–196: Questions

194 **i.** What are the causes of hypercalcaemia in the horse?
ii. Which are the calcium-regulating hormones?
iii. Which diseases of the horse have been reported to be associated with hypercalcaemia of malignancy?

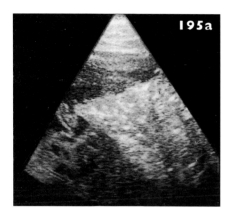
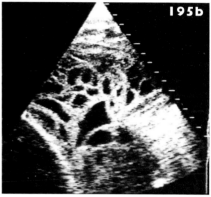

195 You are asked to examine a three-year-old Thoroughbred racehorse who has been spiking fevers up to 40°C and occasionally not finishing its feed.
i. What significant features can you identify in the thoracic sonogram in **195a**?
ii. What is the diagnosis?
iii. What treatment would you recommend?
iv. Over the next several weeks, the horse had intermittent fevers, continued to lose weight and maintained an elevated white cell count. A repeat thoracic ultrasound is illustrated in **195b**? Describe the sonographic findings.

196 A nine-day-old Thoroughbred filly had an abdominocentesis performed with a teat cannula two hours prior to presentation. Now she has a long strand of tissue hanging from the abdomen and the site of the peritoneal tap (**196**).
i. What is the tissue?
ii. What should be done?

194–196: Answers

194 i. Causes of hypercalcaemia in the horse include primary hyperparathyroidism, primary renal disease, hypervitaminosis D (either from dietary supplementation or from plant intoxication) and pseudohyperparathyroidism or hypercalcaemia of malignancy. Renal failure is probably the most common cause of hypercalcaemia in the horse. The equine kidney is a major site of calcium regulation in this species, and renal dysfunction may result in hypercalcaemia and hypophosphataemia. However, this is not a consistent finding in renal failure, and if there is concurrent hypoproteinaemia, hypercalcaemia may not develop, since the majority of blood calcium is protein-bound. Unlike renal disease and hyperparathyroidism, vitamin D toxicosis results in hyperphosphataemia.
ii. Parathyroid hormone (parathormone) has a major role in the regulation of calcium and phosphorus. Parathyroid hormone is an 84-amino acid peptide that is enzymatically cleaved into two fragments. The biological activity resides in the amino-terminal (N-terminal) fragment, but both the N-terminal fragment and the intact molecule have short half-lives. Serum calcium is regulated largely by the interaction of parathyroid hormone and vitamin D metabolites on the bone and gut. Vitamin D can be regarded as a steroid hormone. The two naturally occurring precursor forms of vitamin D, ergocalciferol and cholecalciferol, become vitamin D upon irradiation in the skin. The hormone is subsequently hydroxylated in the liver to form 25-hydroxycholecalciferol. The more bioactive metabolite, 1,25-dihydroxycholecalciferol is synthesized after secondary hydroxylation in the kidney. The hydroxylation is under the influence of parathyroid hormone in response to a decreased plasma ionized calcium concentration. Vitamin D metabolites act on the intestine to promote the uptake of calcium, and together with parathyroid hormone they act on bone to cause resorption of calcium. Calcitonin and parathyroid hormone have an antagonistic effect on bone resorption and a synergistic effect on decreasing the renal tubular reabsorption of phosphorus.
iii. Reports of hypercalcaemia of malignancy in the horse have been associated with lymphosarcoma, squamous cell carcinoma, ameloblastoma, a metastatic carcinoma of endocrine origin and adrenocortical carcinoma.

195 i. Sonogram of the left thorax at the 8th intercostal space. The ventral aspect of the lung is sonolucent and wedge-shaped compatible with consolidation and is surrounded by echogenic pleural fluid.
ii. Pleuropneumonia.
iii. Bacterial pleuropneumonia requires aggressive, broad-spectrum antimicrobial therapy. Frequently both Gram-positive and Gram-negative aerobic and anaerobic bacteria are present. A typical treatment regime might include a penicillin, an aminoglycoside and metronidazole to broaden coverage for anaerobes. Drainage of pleural fluid and attention to nursing care are other areas that need to be addressed.
iv. Multiple loculations are seen in the pleural space. These loculations are most likely fibrous or fibrinous adhesions joining the visceral and parietal pleura.

196 i. This is omentum and it is not rare for this to protrude from the site of an abdominocentesis in a foal if a sufficient size hole is made.
ii. This should be cut with scissors flush with the skin and a 'belly wrap' applied.

197–199: Questions

197 You are called to examine a three-week-old Arabian foal with diarrhoea. Two of the 11 other foals at the farm have had mild diarrhoea that began around 10 days of age and persisted for almost two weeks. Faecal consistency is watery, and the foal is depressed. Physical examination reveals no abnormalities

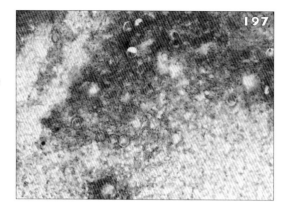

other than diarrhoea and mild dehydration. You submit faecal samples to be cultured for *Salmonella* spp., to be examined for viruses by electron microscopy, and to be tested for clostridial organisms and relevant toxins. In addition, you perform an acid-fast stain of a faecal smear for microscopic examination. Using x100 magnification, you see the image shown in **197**.
i. What is your diagnosis?
ii. What treatment should be recommended?
iii. What steps should be taken for control?

198 i. What is the normal gestational length in the mare?
ii. How do you define prematurity and dysmaturity?
iii. What are the physical characteristics of prematurity in the foal?
iv. How would you treat a premature foal, and what are its chances of survival?

199 A three-year-old Standardbred gelding is examined because of depression, anorexia and sweating. The gelding had tied up after a race 48 hours earlier and had been treated with flunixin meglumine. On examination the gelding had a heart rate of 66bpm, bright red mucous membranes and was restless. Abnormal blood laboratory finding were: PCV, 80%; chloride, 84mEq/l; calcium, 3.7mmol/l (14.8mg/dl); creatinine, 565μmol/l (6.4mg/dl); urea nitrogen, 15.7mmol/l (43.9mg/dl); AST, 40,723u/l; CK, >400,000u/l. The gelding was treated with two litres of 7% NaCl and 15 litres of lactated Ringers solution intravenously. There was no urine produced and the central venous pressure had incresed to 18cm of water (normal <12). What would be the proper treatment for the renal failure in this horse?

197–199: Answers

197 i. Cryptosporidiosis (multiple pink-to-red stained oocysts by acid-fast stain, approximately 4–6 μm in diameter).
ii. No specific treatment exists, although paromomycin and hyperimmune bovine colostrum have been effective in reducing the duration and severity of infection in other species, including calves. Supportive treatment with fluids, nursing care and other medications as needed (eg. anti-inflammatory and antimicrobial drugs).
iii. Isolation of affected foals, especially from younger foals. Careful attention to hygiene, foot baths with appropriate disinfectants (eg. bleach), controlling fomite transmission and wearing protective clothing are important. Cleaning and disinfecting stalls with steam, 10% formalin, 5% ammonia and undiluted commercial bleach is advised; prolonged exposure, which can be difficult to achieve, is required for these liquids. Bedding should be removed and disposed of.

198 i. Normal gestational length varies from 320–360 days. Significant variations exist between horses, and gestational length alone is not a good indicator of readiness for birth.
ii. Prematurity: foal less than 320 days gestation. Dysmaturity: signs of immaturity and/or prematurity in foal more than 320 days gestation.
iii. The physical characteristics of prematurity include small for gestational age; soft or silky hair coat; floppy ears; increased range of joint motion; lax flexor tendons; abnormal progression through normal events subsequent to foaling (eg. standing, suckling); weakness or 'floppiness'; domed forehead; hypothermia; tachypnoea and dyspnoea.
iv. Treatment of premature foals includes the following:
- Supportive care is of primary importance: maintain warm environment (blankets, lamps, heating pads, etc.); general nursing care.
- Respiratory support: intranasal oxygen if hypoxaemic; surfactant treatment if deficient; general nursing care – frequent turning from side to side if recumbent.
- Cardiovascular support if necessary.
- Nutritional support: i/v dextrose; enteral nutrition if GI function is mature; parenteral nutrition if GI function is immature.
- Adrenocortical axis support: the use of corticosteroids is controversial.
- Physical therapy: bandages, splints, braces, etc. as needed for musculoskeletal support; exercise should be limited if hypoplasia of the carpal or tarsal bones is present.
- Prevention of infection: determine adequacy of passive transfer of immunoglobulins; administer colostrum or plasma if necessary; broad-spectrum antimicrobials.

The prognosis depends to a large extent on the gestational age at birth. Survival of foals induced before 320 days is poor. With adequate nursing and intensive care, a 73% survival rate is expected in spontaneous births between 280–322 days gestation. The prognosis is guarded if the neonate does not develop a righting reflex or suckle reflex.

199 Dopamine (5–7μg/kg/min) should be administered intravenously along with frusemide (furosemide) (250mg every four hours) until the horse converts to polyuric renal failure. All other intravenously administered fluids should be discontinued. Both central venous pressure and systemic arterial pressure should be monitored.

200–202: Questions

200 You are asked to examine a nine-month-old Warmblood foal that has developed a soft swelling of the throat (**200**). You are told that the swelling is variable in its size. On palpation the swelling is soft, and does not feel warm or hot. On percussion a tympanitic sound is heard. When gentle pressure is applied, the size of the swelling reduces.

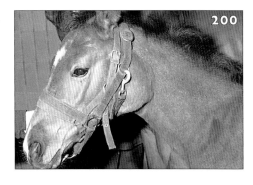

There is no pyrexia, nasal discharge or lymph node enlargement.
i. What is the most likely diagnosis?
ii. What is the cause of this condition?
iii. What other clinical signs may develop?
iv. How would you treat this foal?

201 i. Describe the process of transfer of passive immunity in the newborn foal.
ii. What are the causes of failure of transfer of passive immunity in the foal?

202 This left half of a chest wall (**202a**) came from an eight-month-old foal with a history of a non-healing leg fracture and subsequent jaw fracture. Two of its ribs have been bisected as well. **202b** shows a transected rib from another foal with a similar history, but with different bone fractures. The transection was easily accomplished with a sharp knife.
i. What is the name of this obvious, knobby gross appearance of the costochondral junctions, and what is the name of the distinct white zone in the bone at this joint?
ii. What is the most common cause of this problem?

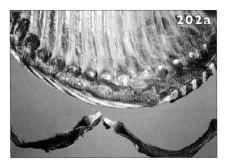

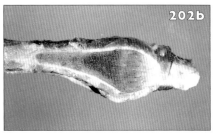

200–202: Answers

200 i. Tympany of the guttural pouch.
ii. The precise cause is uncertain. The plica salpingopharyngea is thought to act as a one-way valve allowing air to enter the guttural pouch during expiration, but preventing it from leaving because of collapse of the pharyngeal orifice as air is expelled. The normal function of this tissue flap may be impeded due to a congenital defect or due to swelling of the flap secondary to upper airway inflammation. In most cases, however, the tissue flap appears to be anatomically normal. Therefore, it is suggested that a functional defect is involved.
iii. In many foals, distension of the guttural pouch is well-tolerated, and there may be no other clinical signs. In severe cases there may be dysphagia, and there is a risk of milk aspiration and the development of inhalation pneumonia. Occasionally, a secondary infection may occur within the distended pouch.
iv. Temporary relief can be obtained by aspiration of air transcutaneously, or installation of an indwelling guttural pouch catheter. Most cases of guttural pouch tympany are unilateral, and fenestration of the median septum between the two pouches is the preferred treatment. Fenestration can be achieved during surgery via Viborg's triangle or by transendoscopic Nd:YAG laser treatment. If the disease is bilateral, fenestration of the median septum must be combined with surgical enlargement of one of the pharyngeal openings.

201 i. Passive transfer of maternal antibodies does not occur transplacentally in the horse. The mare's colostrum contains high concentrations of antibodies (primarily IgG). The foal must ingest colostrum and be able to absorb the antibodies for passive transfer to occur. The foal's gastrointestinal tract is able to absorb proteins intact via specialised villus epithelial cells in the small intestine. Maximum absorption occurs within eight hours following parturition and the specialized epithelial cells are replaced by more mature cells unable to pinocytose proteins within 24–36 hours.
ii. Failure of passive transfer of immunity may arise because of:
- Inadequate colostrum or inadequate concentration of IgG in colostrum: mares that prematurely lactate (>24 hours prior to parturition) tend to have lower IgG concentration in the colostrum; premature parturition – mares foaling prior to 320 days may not have adequate colostral antibodies; mares on fescue pasture during late gestation can have agalactia, thickened fetal membranes, retained placenta, and weak foals.
- Failure of the foal to ingest and absorb colostrum within the first 12 hours. Any illness or abnormality which causes the foal to be weak can result in lack of nursing.

202 i. The rachitic rosary and the osteodystrophic line.
ii. The basic underlying cause is a vitamin D deficiency at the cellular level. One of these two cases was related to keeping the foal inside all day and allowing it out only at night when the flies were not so troublesome. The other case was related to allowing the foal out to exercise every day in a roofed pavilion.

203–205: Questions

203 Over the previous two weeks, on two occasions, an aged retired gelding collapsed to the ground and lay recumbent for periods of up to one hour. Prior to each collapsing episode, it was grazing quietly. The owner reports that while recumbent, the horse lies still but does not appear to have lost consciousness. On physical examination, immediately after the most recent collapsing episode, the main physical findings are a rapid, irregularly irregular pulse rate with a heart rate of 70–80bpm, a loud coarse band-shaped pansystolic murmur loudest over the left fifth intercostal space and radiating caudodorsally, jugular pulsation, moist bronchovesicular sounds and mild ventral oedema. An ECG demonstrated atrial fibrillation. The following day, the horse is

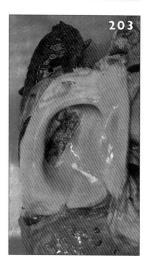

found dead at pasture and a post-mortem examination is performed.
i. The specimen (**203**) illustrates the right ventricular outflow tract and pulmonary artery. What lesion is present?
ii. What is the pathogenesis of this condition?
iii. Given the physical findings in this horse, what other cardiac abnormalities may be present?

204 i. List some alternative names for exertional rhabdomyolysis.
ii. What factors predispose to the development of this disease?
iii. What are the clinical signs of this condition?

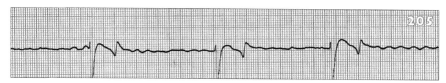

205 A five-year-old Thoroughbred gelding is presented with a history of performing poorly in race training for the previous weeks. The arterial pulses are irregular and vary in strength. On auscultation, there is an irregularly irregular cardiac rhythm with a heart rate of approximately 40bpm but no cardiac murmurs.
i. What does the ECG (**205**) demonstrate?
ii. What is the drug of choice for treatment of this arrhythmia?
iii. What are the main adverse effects of this drug?
iv. What is the prognosis for successful treatment of this condition?

159

203–205: Answers

203 i. There is rupture of the pulmonary artery.
ii. Pulmonary artery rupture usually occurs secondary to chronic pulmonary hypertension. It has been recorded in horses with severe mitral valvular insufficiency and in patent ductus arteriosus. Pulmonary artery rupture can be fatal, or collapsing episodes may precede death.
iii. This horse had a loud murmur with the characteristics of mitral regurgitation; therefore lesions of the mitral valve may be present such as nodular or generalized thickenings associated with degenerative valvular disease, a ruptured chordae tendinae or, less commonly, bacterial endocarditis. With severe mitral valve disease the left ventricle and atrium become dilated and the arrhythmia is most likely to be atrial fibrillation, arising secondary to atrial dilation. Moist bronchovesicular sounds suggest pulmonary oedema and right-sided signs such as jugular pulsation and ventral oedema indicate that biventricular congestive failure is present; therefore, the right cardiac chambers may be enlarged and there may be hepatic congestion, ascites and pleural effusion.

204 i. Exertional rhabdomyolysis is also known as azoturia, 'Monday morning disease', 'tying up', paralytic myoglobinuria and myositis.
ii. There are numerous predisposing factors, acting alone or in combination. The classic description is the draught horse in work that is rested at the weekend and maintained on full high-carbohydrate low-fat feed. When the horse returns to work, it suffers an attack of the disease. A rapid or sudden increase in work intensity or duration also predisposes to the condition, with highly-strung mares and fillies affected more than colts. An increased incidence tends to occur in certain breed lines and families. Respiratory tract viral infections predispose to myopathy in some horses.
iii. The clinical signs vary, depending on the extent of muscle degeneration. Commonly, there is a stiff or stilted gait and reluctance to move. Pain may be elicited by deep palpation of the back and hindlimb musculature. Tachycardia, tachypnoea and pyrexia are sometimes seen.

205 i. The ECG demonstrates atrial fibrillation characterized by absence of P waves, presence of F wave (undulating baseline) and irregular R to R interval.
ii. Quinidine sulphate is the drug of choice.
iii. Quinidine has both cardiovascular and extra-cardiac adverse effects. It is a vagolytic drug which can induce rapid supraventricular tachycardia by increasing conduction through the atrioventricular node. It can induce ventricular arrhythmias. Its alpha-adrenergic action leads to vasodilation and hypotension and, at higher doses, it decreases ventricular contraction. Common extra-cardiac side effects include depression, anorexia, tympanitic colic, gastrointestinal ulceration, diarrhoea, upper airway oedema and respiratory stridor.
iv. The prognosis for successful treatment of the condition depends on the presence of underlying cardiac disease and the duration that the arrhythmia has been present. In this horse, the resting heart rate is normal, there are no cardiac murmurs and the arrhythmia has only been present for two weeks; therefore, the prognosis is good. Echocardiography is the most useful means to assess the presence of underlying cardiac disease prior to treatment.

206 & 207: Questions

206 A two-year-old Thoroughbred mare has suffered two low-grade colic episodes in three days. During these episodes the mare flank watches and paws at the ground, all cardiovascular parameters remain normal and mild small intestinal distension is palpable *per rectum*. Both

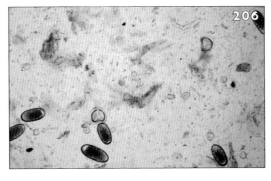

episodes have resolved with medical treatment. As part of your investigation of this case, a faecal sample is taken and a McMaster egg count performed. A result of 200 strongyle epg is obtained. Amongst the strongyle eggs observed under the microscope are the objects illustrated (**206**).
i. What are they, and how may they relate to this animal's problem?
ii. Suggest an appropriate treatment regimen and follow-up diagnostics.

207 A four-year-old Quarterhorse is examined because of recurrent episodes of colic. The episodes reportedly have been occurring every 3–7 days and have been mild in severity. They appear most often after

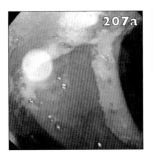
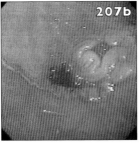

feeding concentrate, and have been increasing in frequency during the previous two weeks. The horse has been used for racing and has been performing intensively. The diet comprises daily feedings of approximately eight pounds of sweet feed, six pounds of oats, and 16 pounds of hay, divided into three feedings. Results of abdominal auscultation, faecal occult blood, parasitological and microbiological examination of faeces, rectal palpation, abdominocentesis, abdominal ultrasonography, haematology and serum biochemistry were within reference ranges. Gastroscopic examination included the images **207a** and **207b** (closer view of same area).
i. Describe the lesions seen in the figures.
ii. What criteria should be used for diagnosis (ie. detection and attributing clinical signs to the lesions)?
iii. What treatment would you recommend?

206 & 207: Answers

206 i. The objects pictured next to the strongyle eggs are tapeworm eggs of the species *Anoplocephala perfoliata*. A strongyle egg count of 200epg indicates a relatively small strongyle worm burden, particularly in a horse of this age, and is probably not significant. The historical and clinical evidence suggests that this mare is suffering from an intermittent, non-strangulating obstruction of the small intestine. A number of different lesions could cause these signs:
- Intussusception (of jejunum, ileum or ileocaecal junction).
- Intermittent intestinal impaction (eg. ileal impaction, ascarid impaction).
- Stenosis of small intestine causing intermittent impactions (eg. adhesions).
- Intermittent entrapment in an internal hernial ring (eg. mesenteric rent).
- Physiological disturbance of intestinal motility ('spasmodic' colic).

The presence of tapeworms in this horse's intestinal tract increases the likelihood of pathology at the ileocaecal junction and the risk of ileocaecal colic.

ii. Pyrantel is the only anthelminthic licensed for use in the horse with proven activity against *A. perfoliata*. Twice the normal nematocidal dose (38mg/kg) is more than 90% effective in removing this parasite from the intestinal tract. Follow-up faecal diagnostics will be useful to confirm the removal of tapeworms. The assay of choice for the identification of tapeworm eggs is faecal centrifugation and floatation using a saturated sugar solution. Continued episodes of colic following treatment for tapeworms is an indication for further investigation, possibly culminating in exploratory laparotomy.

207 i. Gastric ulceration of the stratified squamous epithelium along the lesser curvature at the margo plicatus.

ii. The following criteria should be met for diagnosis:
- Gastroscopic evidence of ulceration.
- Exclude other causes of colic, because ulceration may occur concurrent with other disorders.
- Observe positive response to treatment.
- Repeat gastroscopy after response to treatment for evidence of healing.

iii. Treatment strategies would include use of histamine-2 (H_2) blockers such as ranitidine (4.4–8.8mg/kg p/o bid), antacids, proton pump inhibitors such as omeprazole, and sucralfate. Though effective, omeprazole may be cost-prohibitive in adult horses at current prices. The effectiveness of sucralfate for treating gastric ulcers of the squamous epithelium has been questioned.

208–210: Questions

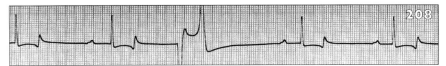

208 A seven-year-old Thoroughbred gelding is pulled up during the cross-country stage of a three-day event when the rider realizes that the horse is extremely tired and is beginning to stagger. Immediately before the competition the horse appeared to be normal; however, four weeks previously the horse had a mild fever for two days which had resolved without treatment. When examined immediately after exercise, the heart was irregular and therefore an ECG (208) was performed the following day.
i. What does the ECG demonstrate?
ii. Whic further investigations are indicated?
iii. What are the treatment options?
iv. How can the response to treatment be assessed?

209 A three-year-old Thoroughbred racing filly had a small corneal abrasion after a race seven days ago. The eye was treated with topical chloromycetin ointment three times a day. It appeared to be improving until yesterday, when marked blepharospasm, lacrimation, photophobia, and ocular cloudiness were observed. You sedate the horse and block the auriculopalpebral nerve for a full examination. The cornea is shown in 209. What diagnosis is most likely?

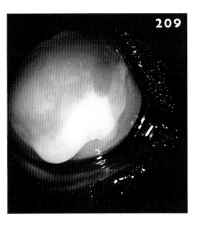

210 This 10-year-old horse has its entire body covered with tightly adherent, painful crusts (210). When the skin disease began, the horse was otherwise normal. Now the horse is depressed, febrile and anorectic. When the crusts are removed under sedation, the subjacent skin is ulcerated.
i. Is dermatophilosis a reasonable differential diagnosis?
ii. Give three differential diagnoses.
iii. What is the most direct method of determining the cause of the condition?

208–210: Answers

208 i. A ventricular premature depolarization (VPD). The third QRS complex is premature and has a different configuration from the other sinus beats, indicating that it originated in the ventricle. In contrast, supraventricular premature depolarizations are also premature but then take a normal path through the ventricle; therefore, the configuration of the QRS complex which follows a supraventricular premature depolarization is normal.

ii. Further investigations would include:
- 24-hour ambulatory ECG is useful to determine the frequency of the arrhythmia.
- Echocardiography may demonstrate ventricular dilation and decreased contractility in some horses with myocardial disease and is used to rule out the existence of lesions such as endocarditis, pericarditis and aorto-cardiac fistula.
- Haematology and serology can provide evidence of bacterial or viral infection.
- Serum concentrations of cardiac isoenzymes (CK–MB; lactate dehydrogenase isoenzymes), although not specific for cardiac disease, are often increased in the early stages of myocarditis.
- Measurement of serum concentrations of electrolytes and arterial blood gases can demonstrate underlying causes of the arrhythmia.

iii. The first priority is to correct any underlying causes such as electrolyte disturbances. With this clinical history it is more likely that there is primary myocardial disease and therefore anti-inflammatory therapy is indicated. Provided there is no viraemia, corticosteroids combined with pasture rest can be successful. This is normally given for 2–4 weeks. Specific anti-arrhythmic therapy with drugs such as lignocaine (lidocaine), quinidine or procainamide is not necessary. This is only indicated if the arrhythmia is life-threatening, if there are clinical signs of low cardiac output such as pallor or weakness, if the heart rate is greater than 100bpm, or if the VPDs are multiform or if the R on T phenomenon is present (VPDs occurring almost simultaneously).

iv. Twenty-four hour ambulatory ECGs are useful to assess the resolution of VPDs. During this procedure a small monitor is attached to a surcingle and the ECG is recorded on magnetic tape for up to 24 hours. The horse can be left unattended, removing environmental and psychological influences. The true frequency of intermittent arrhythmias such as VPDs can be more accurately documented. Rest should be continued until the 24 hours ambulatory ECG is normal. Before the horse is returned to ridden exercise, an exercising ECG should be obtained on a treadmill or by radiotelemetry to ensure that the arrhythmia is not occurring during exercise.

209 *Pseudomonas* spp. infection of the cornea with stromal melting secondary to the potent exotoxins of the organism.

210 i. No. Generalized dermatophilosis is a chronic disease where crusts tend to disappear and the body becomes covered with scale and loosely adherent crusts. The tightly adherent, painful crusts would be more typical of acute dematophilosis, but this is usually localized or regionalized.

ii. Equine exfoliative eosinophilic dermatitis and stomatitis, equine sarcoidosis, pemphigus foliaceus.

iii. Skin biopsy.

211–213: Questions

211 i. What is the aetiology of hyperlipaemia, and how can you confirm it?
ii. How can urea nitrogen aggravate hyperlipaemia?
iii. What are the major clinical signs of equine hyperlipaemia?
iv. In which serum electrophoresis fraction can very low density lipoproteins be found?
v. What two pathways can lead to metabolic acidosis in equine hyperlipaemia?

212 A seven-year-old Thoroughbred mare (**212a**) has developed weight loss and polypnoea during the past two months. The mare was vaccinated against influenza and equine herpesvirus 1 (rhinopneumonitis), and an immune modulator was given within the past three months. On clinical examination, wheezes and crackles are noted on thoracic auscultation. The temperature is normal but both fibrinogen (9.0g/l) and total plasma protein (88g/l) are abnormally high. A transtracheal wash reveals an inflammatory but non-septic exudate. A radiograph is taken (**212b**).
i. What are the radiographic findings?
ii. What are the differential diagnoses?

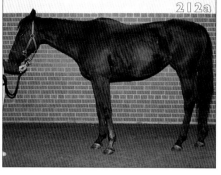

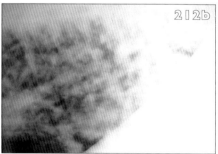

213 The owner of a newborn foal has noticed that one of the foal's eyes looks 'strange' (**213**). What is this lesion?

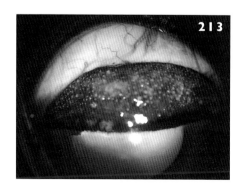

211–213: Answers

211 i. Fatty acids are mobilized from adipose tissues in horses in response to negative energy balance (eg. reduced food intake, pregnancy, lactation or disease), ACTH and glucocorticoids. Although some of these fatty acids are used by peripheral tissues, most are taken up by the liver, where they can be oxidized completely to provide energy or partially to provide ketone bodies (eg. β-hydroxybutyrate). Some fatty acids will be re-esterified to triglycerides and phospholipids, which accumulate in the liver or are released into the circulation as very low density lipoproteins. The very low density lipoproteins are the primary lipoproteins that are elevated in the blood of hyperlipaemic ponies. Hypoglycaemia decreases insulin release, thereby enhancing lipid mobilization from adipose tissue and decreasing triglyceride removal from the peripheral blood. Although insulin resistance has been proposed as the likely cause of the hyperlipaemia, it may be accompanied by normal insulin responses. Obesity has been associated with insulin resistance in ponies.

The diagnosis can be made by visual inspection of plasma (**211**) and demonstration of increased plasma lipid concentration, sometimes accompanied by hypoglycaemia. In addition, liver enzyme activities are usually elevated due to hepatic lipidosis.
ii. Urea nitrogen inhibits lipoprotein lipase, resulting in a decreased uptake of triglycerides by peripheral tissues.
iii. The major clinical signs seen in these horses are anorexia and lethargy. Other clinical signs may include diarrhoea, ventral oedema and icterus.
iv. The β-globulin fraction.
v. Lactic acid is buffered by HCO_3^-, resulting in the formation of lactate. Lactate is metabolized, primarily in the liver but in the kidneys as well, to form pyruvate, which is then converted into either CO_2 and H_2O (catalyzed by pyruvate dehydrogenase) or glucose (catalyzed by pyruvate carboxylase). Both pathways result in the formation of HCO_3^- as well. Renal and hepatic lipidosis reduce the utilization of lactate, resulting in reduced HCO_3^- formation and the accumulation of excess lactic acid.

Metabolic acidosis can also result from diminished net tubular H^+ secretion due to renal lipidosis, as well as to hypovolaemia and increased lactic acid production. Horses suffering from hyperlipaemia in combination with a plasma pH below 7.15 and/or a plasma triglyceride concentration >1,200mg/dl have a very poor prognosis.

212 i. There are multifocal opacities in the lung.
ii. The differential list would include neoplasia, mycotic pneumonia, haematogenous bacterial pneumonia and, as in this case, granulomatous pneumonia.

213 Congenital dermoid. These are misplaced sections of haired skin. They may appear anywhere on the eye, lids or conjunctiva. However, they are most common at the limbal region of the globe. The hairs associated with equine dermoids are very small and can be missed unless magnification is used.

214 & 215: Questions

214 A 16-year-old pregnant (three months) Warmblood mare is presented with a history of low grade colic of five days duration. For the last three days she has been anorexic. The major clinical features are weight loss and ventral oedema (**214a**). Palpation per rectum reveals enlarged mesenteric lymph nodes. Haematological

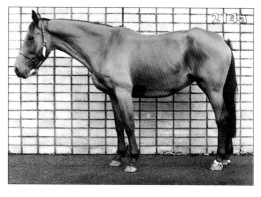

evaluation reveals a haematocrit of 49% (n, 36–42%), and leucopenia (WBC count, $4.1 \times 10^9/l$; n, $7–10 \times 10^9/l$) with 26% lymphocytes and 74% neutrophils. The total serum protein and albumin concentrations are low (total protein, 30g/l; n, 60–80g/l; albumin, 12g/l; n, 35–40g/l). Your tentative diagnosis is alimentary lymphosarcoma.
i. What further blood biochemical analyses or tests may be helpful in supporting your diagnosis?
ii. What would be your tentative diagnosis if soft tissue mineralization of the aortic wall is found at necropsy in a horse?

215 A two-year-old Thoroughbred colt (**215**) is examined for ataxia of all four limbs of at least two weeks duration. The colt was being readied for training when the ataxia was noted. There is no known history of trauma. The ataxia is graded 2/4 in the hindlimbs, 1/4 in the forelimbs and seems nearly symmetric. There are no signs of brainstem or

cerebral dysfunction nor is there any evidence of muscle atrophy.
i. What are the three most likely differential diagnoses?
ii. What factors are supportive of cervical compressive myelopathy?
iii. What diagnostic tests should be performed?

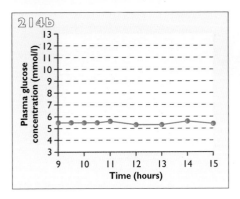

214 i. The ante-mortem diagnosis of alimentary lymphosarcoma can be very challenging. The oral glucose tolerance (absorption) test can be used to demonstrate impaired small intestinal glucose absorption. This test is performed by administering glucose at a dose of 1g/kg bodyweight by nasogastric tube after withholding feed for 12 hours. Blood samples are collected for plasma glucose estimation at 0, 30, 60, 90, 120, 180, 240, 300 and 360 minutes. Normal small intestinal absorption produces a peak plasma glucose at 120 minutes of 200% of basal level, and return to basal levels by 360 minutes. This mare demonstrated only a 1.8% increase in plasma glucose (**213b**), indicating small intestinal malabsorption. Common causes of malabsorption include alimentary lymphosarcoma, inflammatory bowel diseases (such as granulomatous enteritis, diffuse chronic eosinophilic enteritis, lymphocytic enteritis), equine motor neuron disease and extensive small intestinal resection.

Hypercalcaemia is sometimes found in horses affected by neoplastic processes, including lymphosarcoma. Other biochemical abnormalities that are sometimes present in cases of alimentary lymphosarcoma include low serum IgM levels and raised serum alkaline phosphatase levels.

ii. The tentative diagnosis (in the absence of any evidence of malignancy) would be hypervitaminosis D. The disease may result from overdoses of injectable or oral supplements, or ingestion of toxic plants with vitamin D activity. Serum calcium itself is an unreliable indicator of vitamin D toxicosis as it fluctuates during the course of the disease.

215 i. Cervical compressive myelopathy, equine degenerative myeloencephalopathy and equine protozoal myeloencephalitis (EPM) (*Sarcocystis neurona* or *falcatula*), if the horse originated from an endemic area.
ii. The breed, sex, age and clinical signs are all supportive of cervical compressive myelopathy.
iii. Spinal fluid should be obtained and tested for EPM (which was positive in this case). If EPM is ruled out, radiographs of the cervical spine should be performed. A myelogram will be required to confirm a cervical compressive lesion. A slap test will be of little benefit.

216–218: Questions

216 A three-year-old Thoroughbred racehorse falls and is unable to stand or right himself into a sternal position. A spinal tap is performed in the lumbosacral area and fluid is collected which is uniformly discoloured (**216**).
i. What is the prognosis?
ii. Where is the neuroanatomical location of the lesion that would best explain the clinical signs?
iii. How could it be determined whether the discolouration in the CSF is a result of the disease or iatrogenic from the tap?

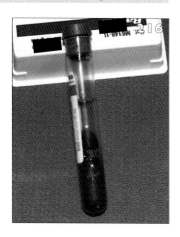

217 Shown (**217a**, **217b**) are post-mortem appearances of the ventral abdominal wall of an adult, cross-bred horse which was sent to necropsy because of a fractured femur. The only remarkable lesions found at necropsy, other than the fractured femur, were these almost symmetrical, bilateral areas of firm tissue in the retroperitoneal fat along both sides of the linea alba (**217a**), which on cross-section showed a mixture of fat and connective tissue (**217b**). What is this process called, and what is its significance?

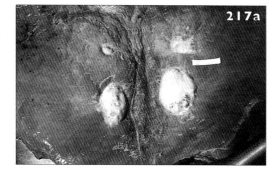

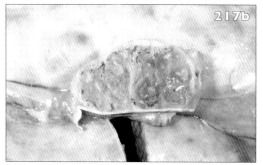

218 i. Describe the usual clinical course of strangles.
ii. How would you normally treat strangles?
iii. What potential complications may develop from this disease?

216–218: Answers

216 i. The prognosis is poor to grave. Large amounts of blood appearing uniformly in the spinal fluid of a recumbent horse, known to have experienced trauma, generally indicates a vertebral fracture. This horse was confirmed as having three vertebral fractures.
ii. Since the horse cannot gain a sternal position, the cord compression is most likely in the cervical vertebrae.
iii. If the colour of the CSF does not change during the collection period, it suggests that the changes are not iatrogenic. Cytologically, the absence of platelets is also not supportive of a traumatic tap.

217 These areas are called bilateral retroperitoneal abdominal fat necrosis of horses. These are not of any clinical significance, but they are common in horses, especially small breeds. Their cause is unknown.

218 i. Strangles is commonest and most severe in young horses (less than four years of age). After an incubation period of 3–14 days, the initial signs include depression, pyrexia (39.0–40.0°C), a mucoid nasal discharge and slight cough. The nasal discharge rapidly becomes mucopurulent or purulent, and the intermandibular area swells and becomes tender to the touch. The submandibular lymph nodes enlarge and become painful, so a horse may stand with its head and neck extended. The swollen lymph nodes start to fluctuate as abscesses develop, and they usually rupture 7–14 days after the onset of clinical signs. Enlargement of the retropharyngeal lymph nodes sometimes causes respiratory distress or dysphagia. Following rupture and drainage of the abscesses, most horses make a complete recovery, and the duration of clinical signs may be less than one week to more than two months. In older animals or 'partially immune' horses, a milder 'atypical' form of disease may occur; abscesses develop in only a small proportion of such cases. Mild 'atypical' strangles has also been associated with particular strains of *Str. equi*.
ii. Treatment of uncomplicated cases of strangles is usually symptomatic. Procaine penicillin G is effective in treating the infection, but treated animals are unlikely to mount a protective immune response and they may be susceptible to further infection. Where abscesses have developed, antibiotic therapy may delay their maturation or drainage. It has been suggested that antibiotic treatment might also increase the risk of metastatic spread of the organism to other parts of the body. Hot fomentation or poulticing of abscesses is helpful. Surgical drainage of an abscess is occasionally necessary if spontaneous rupture fails to occur. Tracheotomy should be considered in cases where there is severe respiratory distress due to upper airway obstruction. Measures should be taken to limit the spread of infection to other susceptible horses.
iii. Potential complications include cellulitis; empyema of the guttural pouch; pneumonia; abscess formation in other parts of the body, eg. thoracic or abdominal lymph nodes, brain, liver, kidney, spleen ('bastard strangles'); pleuritis; peritonitis; endocarditis; myocarditis; tenosynovitis; mastitis; spinal or vertebral body abscessation; purpura haemorrhagica; anaemia.

219–221: Questions

219 This post-mortem view (**219**) demonstrates a sinonasal adenocarcinoma in an aged Thoroughbred mare. This mare presented with a complaint of epistaxis. The mare had been diagnosed initially as having immune-mediated thrombocytopenia and treated with dexamethasone. Although the mare's platelet count increased with this treatment, it did not

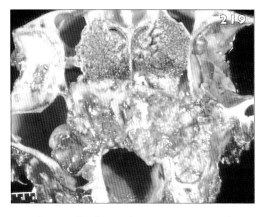

come close to a normal range. Upon thorough physical examination and diagnostic evaluation, a space occupying mass was identified in the sinus.
i. How do you explain the thrombocytopenia in this mare with primary sinus neoplasia?
ii. Is there any other possible explanation for the relatively non-responsive thrombocytopenia in this mare?

220 i. What tests are commonly used to diagnose failure of passive transfer of immunity in a foal?
ii. What treatments are available for failure of passive transfer of immunity?

221 Shown (**221**) is the left eye of a five-year-old Quarterhorse that appeared to have mild blepharospasm, lacrimation and photophobia seven days ago. The owner treated the eye with some ophthalmic ointment and the pain seemed to lessen for a few days. However, over the previous two days, the eye has become increasingly painful and cloudy. You sedate the horse, block the auriculopalpebral nerve and find that the eye is as illustrated. Fluorescein dye test is negative. What is your diagnosis?

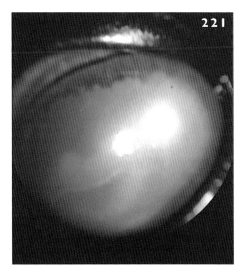

219–221: Answers

219 i. It is possible that a tumour related antigen was labelling the mare's platelets, causing an autoimmune destruction of thrombocytes.
ii. Yes. This mare had pseudothrombocytopenia. All her platelet counts were performed on blood collected into EDTA. When a platelet count was performed on heparinized blood, the platelet count was found to be normal. There is a specific phenomenon, that occurs in certain horses and people, in which platelets react with EDTA resulting in a falsely low platelet count. Whenever thrombocytopenia (<40,000/µl) is diagnosed in a horse without clinical signs of petechiae, the platelet count should be repeated on blood collected into heparin as an anti-coagulant.

220 i. Tests for measuring immunoglobulin levels in the foal should be performed at 24 hours of age (after antibody absorption is complete). Serum IgG values <400mg/dl are considered diagnostic of failure of passive transfer of immunity; 400–800mg/dl is considered partial failure; >800mg/dl is considered normal.

The zinc sulphate turbidity test is the least expensive test available and results are available within one hour. However, this is not as sensitive as other tests and false positives can occur.

The ELISA test is an immunoassay. Results are obtained rapidly and are fairly reliable. However, this is an expensive test.

Radial immunodiffusion (RID) is the most accurate test available, but the results take 18–24 hours.
ii. The treatment is dependent on the reason for failure of passive transfer of immunity and the age of the foal.

If the foal is <18–24 hours old: frozen colostrum can be administered via bottle or nasogastric tube. At least one litre of colostrum should be administered, preferably within the first 12 hours of life, divided into several feedings. Frozen colostrum should not be microwaved. Lyophilized IgG is an oral IgG supplement, but variable absorption has been reported so the IgG levels should be measured. Plasma can be administered orally, but it is not considered to be as effective as intravenous administration.

If the foal is >24 hours old or if colostrum is not available: plasma transfusion. Approximately 20–40ml/kg plasma is necessary to raise IgG levels to >400mg/dl. If the foal is already sick, additional amounts will be necessary. Ideally, the plasma should be cross-matched if commercial cell-free plasma from anti-RBC negative donors is not being used. Serum IgG levels should be tested at least 1–3 hours following administration and additional plasma could be administered if the IgG levels are still low. Caution should be taken not to cause fluid overload, especially in sick or very small foals.

221 Corneal stromal abscess. These lesions frequently do not retain fluorescein dye since the corneal epithelium may have already healed over the initiating small corneal puncture that inoculated the stroma with infectious organisms.

226–228: Questions

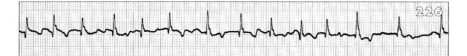

226 A three-year-old Highland pony is presented with swelling of the sheath and muzzle, ventral oedema and lethargy for the previous two weeks. The arterial pulses are irregular and vary in strength. On auscultation there is an irregularly-irregular cardiac rhythm with a heart rate of approximately 80bpm, a coarse grade 6/6 band-shaped systolic murmur loudest in the right fourth intercostal space and a grade 4/6 crescendo-decrescendo systolic murmur loudest in the left third intercostal space.
i. What does the ECG demonstrate (**226**)?
ii. Which cardiac conditions cause a murmur loudest over the right hemithorax?
iii. How should this case be managed?

227 Abdominocentesis (**227**) is a useful investigative technique in certain acute and chronic colic problems. The indications for this procedure include the suspicion of early strangulating obstruction, non-strangulating infarction, peritonitis and intra-abdominal neoplasia.
i. What are the potential complications of performing abdominocentesis and the (relative) contraindications?
ii. How may the risk of complications be minimized?

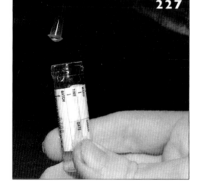

228 You are asked to examine a four-year-old gelding with a history of poor performance and intermittent nasal discharge of food material. Clinical examination reveals no abnormalities, and you decide to perform an endoscopic examination (**228**).
i. What is your diagnosis?
ii. What is the cause of this condition?
iii. What are the therapeutic possibilities?
iv. What is the prognosis?

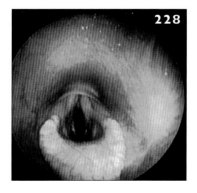

226–228: Answers

226 i. The ECG demonstrates atrial fibrillation.
ii. Tricuspid insufficiency and ventricular septal defect are the principal lesions which cause a murmur loudest over the right hemithorax. In horses a ventricular septal defect usually causes a relative pulmonic stenosis and, therefore, there is also a murmur loudest in the left third intercostal space.
iii. In this horse the presence of tachycardia, the clinical signs of congestive heart failure and the loud cardiac murmurs indicate that atrial fibrillation has arisen secondary to significant underlying cardiac disease. An echocardiogram confirmed the presence of a very large ventricular septal defect. In contrast to the horse described in 205, quinidine sulphate is contraindicated in this horse. It will not address the underlying lesion, and its vagolytic and alpha-adrenergic effects can be fatal in horses in congestive heart failure. Short-term alleviation of signs may be achieved with the diuretic frusemide (furosemide) and digoxin, which will decrease the heart rate. The long-term prognosis is hopeless.

227 i. A sharp needle introduced into the abdominal cavity has the potential to perforate or tear bowel or other organs. It is speculated that the normally motile intestine will move away from a sharp point and contract around penetrations to seal them. This is consistent with the finding that problems relating to this technique are most commonly seen when intestinal motility is compromised or when the intestine is held against the ventral abdominal wall. Situations in which abdominocentesis carries a greater than usual risk of iatrogenic damage are:
- Intestinal distension (small or large intestine) (mainly in foals).
- Heavily gravid uterus.
- Adhesions between intestine (often caecum) and ventral abdominal wall.

ii. The first two of these risk factors are often identifiable on rectal examination, so it is advisable to perform abdominocentesis only **after** rectal examination. The use of a blunt teat cannula may reduce the risk of inadvertent bowel puncture but is a more invasive procedure as it requires a scalpel cut through the skin and linea alba.

228 i. Rostral displacement of the palatopharyngeal arch (RDPPA). The palatopharyngeal arch is seen covering the apical area of the corniculate cartilages, producing the appearance of a hood of tissue over the top of the larynx.
ii. There has been an abnormal development of the fourth branchial arch resulting in an abnormal anatomical development of the thyroid cartilage. This causes displacement of the caudal margin of the ostium intrapharyngeum. There is impaired articulation of the thyroid cartilage with the cricoid cartilage and limitation of the movements of the arytenoid cartilages.
iii. Surgical removal of the palatopharyngeal arch through a ventral laryngotomy is indicated when there is no dysphagia and the only problem the animal appears to have is obstruction of the airways.
iv. The prognosis depends on the severity of the developmental abnormalities, but is usually guarded to grave. Even in animals where there is only partial obstruction of the airways, there is always a degree of anatomical abnormality that will impair upper airway function even after successful surgery.

229–232: Questions

229 i. What is the pathogenesis of tetanus?
ii. What is the prognosis?

230 A newborn Thoroughbred foal seems to have poor vision according to the owner, who reports that one of the foal's eyes appears cloudy but the other appears normal. With a penlight, you find that the left eye has a congenital cataract, but the right eye does not. Focal light examination of the right eye reveals a fixed dilated pupil, and fundic examination looks like **230**. What is your diagnosis?

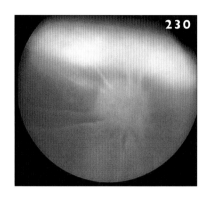

231 A nine-year-old Thoroughbred gelding is examined because of stiff gait for more than one year and urinary incontinence for at least one month. The gelding exhibits stranguria, pollikauria and intermittently dribbles urine while either standing in the stall or at a walk (**231**). Rectal examination reveals a very large bladder that cannot be easily expressed. A catheterized urine sample reveals large numbers of red blood cells and white blood cells but no bacteria are observed or cultured. After catheterization, re-examination of the bladder reveals a large amorphous 'gritty'-feeling mass in the bladder lumen. Repeat catheterization of the bladder and lavage with saline is successful in removing a large amount of sand-like sediment from the bladder. A third rectal examination reveals that the mass is no longer present and the bladder is now small but feels as though it has a thickened wall.

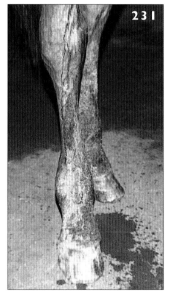

i. What is the term for the sand-like mass found in the bladder?
ii. Why does this form and, in this case, could it be related to the stiff gait?

232 What supportive treatment would be needed for the mare described in **193** and **225**?

229–232: Answers

229 i. Tetanus is caused by intoxication by neurotoxins produced by *Clostridium tetani*. The organism produces spores and grows in anaerobic sites to produce its toxins. The disease has a world-wide distribution, and *Cl. tetani* spores are found in the soil, especially soils heavily contaminated by faecal matter. *Cl. tetani* neurotoxin interferes with interneurons in the CNS resulting in hypertonia and hyperreflexia. Affected interneurons are often inhibitory, therefore there is a lack of inhibition and, consequently, stimulation of lower motor neuron function. Toxin is bound to gangliosides in the CNS, and the effects wear off as these are replaced.

The most frequent sites of infection are soft tissue injuries (including surgery) and puncture wounds to the foot. Tetanus is also seen in young foals through umbilical cord infection and in mares after foaling. Under anaerobic conditions (especially in the presence of necrotic tissue and pus), *Cl. tetani* spores germinate into the vegetative form, which produces toxins. At least three toxins are produced. Tetanolysin increases the local tissue necrosis. Tetanospasmin binds to nerves and is transported to the CNS where it inhibits the action of inhibitory interneurons in the ventral horn of the spinal cord. Non-spasmogenic toxin causes overstimulation of the sympathetic nervous system.

ii. The prognosis is poor. If the horse can still drink, the prognosis is good with nursing care. For recumbent horses, the mortality rate approaches 80%. Horses that survive for more than seven days have a fair chance of survival. Recovery from tetanus does not protect against the disease; therefore, the horse must still be vaccinated. Other horses at risk should be vaccinated. Decubital lesions, fractured bones and scoliosis can be lethal complications of the disease.

230 Retinal detachment. Multiple congenital anomalies are not unusual in foals. They may include combinations of microphthalmia, and lens, retinal and anterior segment pathology.

231 i. Sabulous urolithiasis.
ii. This sediment is a normal component of equine urine that accumulates in the bladder when the bladder cannot be completely emptied. The large sediment mass develops over several weeks or months. Yes, the stiff gait and sabulous urolith are likely to be related. The horse was found on autopsy to have severe bridging spondylitis. The spondylitis may have caused the stiff gait and prevented the horse from attaining the proper stance required to completely void urine from the bladder.

232 Treatment should include intravenously administered polyionic fluids with 20–40mEq/l of KCl and added dextrose to make the final concentration 5–10%. Neomycin (5–10mg/kg) should be given via dose syringe p/o tid for two days. The mare should be fed small amounts of high carbohydrate, high branch chain amino acid feeds (corn or sorghum with molasses) frequently along with oat hay. She should be kept out of the sunlight, but allowed to graze some grass, if it is not lush spring grass which is too high in protein.

222 & 223: Questions

222 You are asked to examine an 11-year-old Thoroughbred brood mare with a history of fever and a slightly haemorrhagic nasal discharge. Lung auscultation was abnormal with a change in the character of breath sounds heard ventrally. Access to thoracic ultrasonography was not available.

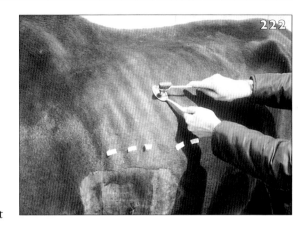

i. What is the name of the technique illustrated (**222**)?
ii. What are the landmarks/borders of the normal horse's lung field?
iii. Although a rubber reflex hammer and ordinary teaspoon work well, what are the names of the specially designed instruments used for this procedure?
iv. If a large volume of pleural fluid is present in this horse, how would you describe the sounds heard during this procedure?

223 An exploratory laparotomy was performed on a nine-month-old Thoroughbred colt which had moderately severe abdominal pain which was minimally responsive to analgesics. The findings at surgery are illustrated in **223**. Two similar cases have occurred in the same batch of 20 weanlings which were born and reared on the same stud farm.

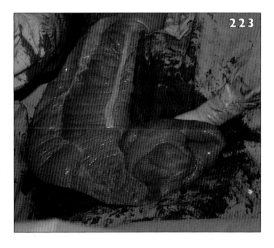

i. What is the diagnosis?
ii. Given the occurrence of several cases, what is the possible aetiology of this condition?
iii. What treatment and management is appropriate for the rest of the group of yearlings?

222 & 223: Answers

222 i. Thoracic percussion.
ii. (a) Cranial – shoulder musculature.
(b) Dorsal – back musculature.
(c) Caudoventral – level with tuber coxae at 17th intercostal space; level with tuber ischii at the 15th–16th; level with the middle of the thorax at the 13th space; level with the point of the shoulder at the 11th intercostal space; the field then extends cranioventrally in a curving line to the point of the elbow.
iii. Plexor (hammer) and pleximeter (flat blade).
iv. Percussion of the thorax of a horse with a large volume of pleural fluid would produce a dullness or flatness of sounds as pleural fluid interferes with the production of normal resonant vibrations. While this same dullness might be heard over consolidated lung, the change from resonant to non-resonant is more obvious when due to pleural fluid.

223 i. Caecocaecal intussusception.
ii. Cyathostome infection. Caecocolic and caecocaecal intussusceptions have been diagnosed in weanling Thoroughbred animals with heavy infections of mucosal stages of cyathostomes (small strongyles). In affected cases there is typically marked oedema, and thickening and inflammation of the caecum and large colon. In addition, there are grossly enlarged, reactive local lymphatics. On close examination there is evidence of coiled cyathostome larvae within the mucosa; these can be identified readily using transillumination with a strong light source and/or magnification. Following surgical correction of the intestinal obstruction these cases have a reasonable prognosis for full recovery to fulfil their athletic potential.
iii. Anthelmintics at 'larvicidal' dosage rates. It is possible that killing maturing larvae and/or adult cyathostomes may provoke colic episodes consequent upon altered intestinal motility. Certainly, prior to administration of anthelmintics, CBCs with differential counts and blood biochemistry analyses for total protein, albumin and globulin, and alkaline phosphatase should be undertaken in all animals in the group. On the basis of leucocytosis, neutrophilia, hypoalbuminaemia, hyperglobulinaemia and raised alkaline phosphatase it should be possible to identify any particularly 'high-risk' individual animals which should be given increased clinical monitoring following treatment. The choice of anthelmintic regimen is complex:
- Many cyathostome populations are benzimidazole resistant, but in most countries there is a label claim for efficacy against mucosal cyathostomes using oral benzimidazoles at either 7.5–10mg/kg bodyweight fenbendazole for five consecutive days, or 30–60mg/kg fenbendazole or 10mg/kg oxfendazole on a single occasion.
- Ivermectin (0.2mg/kg bodyweight repeated at 10 day intervals) has been reported to work well in clinical cyathostome-associated disease, but from controlled drug efficacy studies it is apparent that this drug has only limited efficacy against mucosal cyathostomes.
- Moxidectin is not currently licensed for horses in the US or Europe but this compound has been shown in trials to have fairly good efficacy against mucosal stages of cyathostomes.

224 & 225: Questions

224 You are asked to examine a group of Thoroughbred horses which have developed signs of respiratory infection – harsh dry coughing, pyrexia (up to 41.7°C), lethargy, inappetence, tenderness of submandibular lymph nodes and nasal discharge, which is initially serous but later becomes mucopurulent (**224**). The disease has spread rapidly through the horses, and the clinical signs are most severe in the youngest animals (yearlings and two year olds) and in horses that have not been vaccinated against influenza. Horses in neighbouring yards have also been affected by the disease.

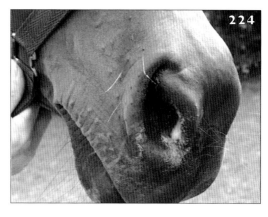

i. What is the most likely virus involved in this outbreak?
ii. What is the pathogenesis of this infection?
iii. How quickly would you expect the horses to recover, and what complications might occur?
iv. How could you confirm the diagnosis?
v. How would you treat these horses?

225 Illustrated here (**225**) is blood from the mare in **193** in a microhaematocrit tube.
i. Which other laboratory tests would you request in order to confirm your suspicion as to the reason for the clinical findings?
ii. What other historical information would you like to have on this mare which might explain the reason for the disorder?

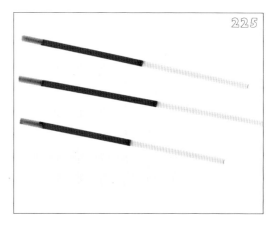

224 & 225: Answers

224 i. Equine influenza virus.
ii. Aerosolized virus is inhaled and deposits on the mucosa of the upper and lower respiratory tracts. The virus attaches to the epithelial cells and enters the cell cytoplasm where replication occurs. The epithelial lining of the entire respiratory tract is affected. Infected epithelial cells are damaged, leading to inflammation, clumping of cilia and focal erosions. Secondary bacterial infections are common.
iii. In uncomplicated cases, recovery occurs in 1–3 weeks. Secondary bacterial infection of the upper respiratory tract and paranasal sinuses can prolong the course of the disease. In a small number of cases, secondary bacterial infection of the lower respiratory tract may give rise to bronchopneumonia with signs of dyspnoea, chest pain and reluctance to move. In a minority of cases, myocarditis occurs causing tachycardia, arrhythmia and severe exercise intolerance.
iv. A presumptive diagnosis may be made on the basis of the clinical signs and rapid spread of disease. The following tests can be used to help confirm the diagnosis:
- Haematology – anaemia, leucopenia and lymphopenia are seen early in the course of the infection (1–5 days); neutrophilia often occurs later as secondary bacterial infections arise; plasma fibrinogen and plasma viscosity may be elevated.
- Virus isolation from nasopharyngeal swabs.
- Serology – acute and convalescent serum samples demonstrate antibody rise.

v. Treatment along the following lines is indicated:
- Complete rest for a minimum of 3–4 weeks in clean, minimum-dust environment.
- General nursing care and provision of palatable food.
- Antibiotic (penicillin or trimethoprim/sulphonamide) treatment is necessary only if there is significant secondary bacterial infection. If bronchopneumonia is suspected, antibiotic selection should be based on culture of transtracheal aspirate.
- Non-steroidal anti-inflammatory drugs such as phenylbutazone are helpful in horses with high fever, depression or muscle stiffness.
- Immunostimulants, such as mycobacterial cell wall extracts, are reported by the manufacturers to be beneficial.
- Bronchodilators and mucolytics may be helpful in some cases.

225 i. Tests to determine the presence or absence of liver disease and dysfunction should be performed. These should include gamma glutamyl transferase, 206u/l (normal, 10–59u/l); aspartate aminotransferase, 2135u/l (normal, 193–450u/l); total bilirubin, 186μmol/l [10.9mg/dl; normal, 5.1–59.8μmol/l (0.3–3.5mg/dl)]; conjugated bilirubin, 32.5μmol/l [1.9mg/dl; normal, <17.1μmol/l (<1.0mg/dl)]; glucose, 6.3mmol/l [113mg/dl; normal, 4.8–6.2mmol/l (86.4–111mg/dl)]; and bile acids,170μmol/l (normal = <12μmol/l or <20μmol/l in an anorexic horse).
ii. Additional tests that may be useful include measurement of prothrombin and partial thromboplastin time and blood ammonia. The historical information that was helpful in this case was that the mare had been vaccinated with tetanus antitoxin at foaling then causing a serum hepatitis (Theiler's disease).

233 & 234: Questions

233 **i.** Describe the clinical signs of uncomplicated equine Cushing's disease?
ii. What are the two major complications of pituitary pars intermedia adenoma?
iii. The dexamethasone suppression test has the risk of inducing laminitis. What could be the possible mechanism?
iv. What else can cause hirsutism?

234 Impactions of the pelvic flexure are one of the more regularly encountered colic types in general practice as demonstrated by the pie-chart in **234**. Diagnosis is based upon characteristic findings on rectal examination: impacted food material palpable within the pelvic flexure which is located in the caudal abdomen or pelvic canal. Such cases carry a good prognosis with medical treatment but can take several days to resolve. Describe how you would manage a case of pelvic flexure impaction and justify your choice of drugs.

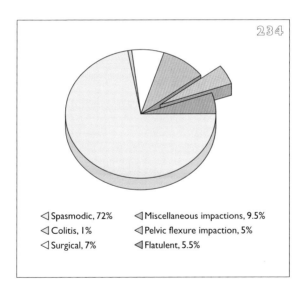

233 & 234: Answers

233 i. The clinical signs of uncomplicated equine Cushing's disease are hirsutism (an excessively long and curly coat – often with a history of failure to shed), hyperhydrosis, good bodily condition, polyphagia and, in about 15% of cases, bulging of the supraorbital fat.
ii. An adenoma of the pars intermedia is the most common disorder of the equine hypophysis. Aged animals are most often affected. The two major sequelae are type 2 diabetes mellitus and laminitis, in 38% and 24% of cases respectively. Other complications include secondary infections (like sinusitis), infertility, pseudolactation, delayed wound healing, blindness due to compression of the optic chiasm, seizures and diabetes insipidus. Pulmonary infection resulting from a mycotic opportunist is possible.
iii. The hoof vasculature is under the control of the sympathetic nervous system via both neurogenic (norepinephrine) and humoral (epinephrine from the adrenal medulla) mechanisms. Corticosteroids create venous obstruction in the distal part of the limb by enhancing the venoconstrictor potency of epinephrine, norepinephrine and serotonin. With regard to this mechanism, a synthetic corticosteroid such as betamethasone is more potent than cortisol itself in enhancing catecholamine-induced venoconstriction.
iv. Besides equine Cushing's disease, long-term anabolic steroid treatment is reported to cause hirsutism. In addition, hirsutism has also been associated with a meningioma located in the pituitary gland.

234 Perhaps the most important aspect of case management is the removal of all food from the horse's environment until the impaction has cleared. Free access to drinking water must be maintained. Therapy is aimed at controlling pain and softening the impaction with intestinal lubricants. An important concept to appreciate is that the veterinary surgeon treating the horse is reliant upon natural gut motility to mix the lubricant in with the food material and then to move the softened material aborally along the intestinal tract. For both these reasons it is desirable that drugs used for pain relief have minimal inhibitory effect upon colonic motility. The non-steroidal anti-inflammatory drugs (NSAIDs) are usually the first choice for pain relief in cases of pelvic flexure impaction. They have two major advantages over other drugs available:
- Prolonged duration of action. This is a desirable characteristic as the pain is likely to last for more than an hour or two.
- Minimal inhibition of gut motility. There is some evidence that phenylbutazone even causes increased motility by inhibiting the 'prostaglandin brake' on gut motility.

Fluids, with or without the addition of intestinal lubricants, should be given by nasogastric tube. Liquid paraffin or mineral oil are the most commonly used intestinal lubricants; a safe volume to give an average 500kg horse is approximately four litres. This dose may be repeated once or twice at intervals of 8–12 hours. A further adjunct to the treatment already outlined is the use of oral and/or intravenous fluids. In the field situation this may be difficult and in many less severe cases it will prove unnecessary. However, horses with particularly large, firm impactions or those slow to respond to initial treatment will often benefit from the administration of isotonic fluids. Magnesium sulphate solution (Epsom salts) administered by nasogastric tube, with or without intestinal lubricants, is another effective laxative treatment.

235 & 236: Questions

235 You are asked to examine a four-year-old Thoroughbred gelding that has been affected by acute and extensive bilateral epistaxis that occurred during the night. The bleeding has now stopped. The animal had not been exercised before the onset of the epistaxis, and there was no history of prior respiratory disease. Endoscopic examination is performed (**235**).
i. What is your diagnosis?
ii. How would you treat this horse?

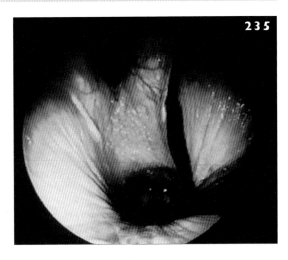

236 This lesion (**236**, area has been clipped) was noticed at a prepurchase examination of a five-year-old horse. The owner reports that it has been present for approximately 12 months and has remained unchanged.
i. What is the most likely diagnosis?
ii. The new owner wants the lesion surgically removed. What problems might arise with surgical intervention?
iii. What new mode of treatment is reported to be 100% effective?

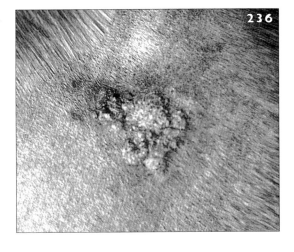

235 & 236: Answers

235 i. Guttural pouch haemorrhage, probably secondary to guttural pouch mycosis. The endoscopic view (**235**) shows a stream of blood emerging from the pharyngeal opening of the auditory tube.
ii. Once a haemorrhage has started, treatment should be aimed at minimizing the risk of a fatal haemorrhage. This entails surgical ligation of the internal carotid artery. There may be back flow of blood from the circle of Willis to the damaged part of the internal carotid artery, so balloon-catheter occlusion of this artery in combination with ligation close to the common carotid artery is a safer procedure. Depending on the degree of blood loss, a blood transfusion may be required. In many cases, administration of saline or Ringer's solution is sufficient. Blood and diphtheritic membranes can be surgically removed from the guttural pouch, but sometimes flushing of the pouch with large amounts of warm saline through a catheter removes most of the debris. Antifungal treatment is best performed topically through an indwelling catheter, inserted through the nose or during surgery. Thiabendazole and enilconazole have been successfully used for this purpose. The response to this treatment, however, is slow and treatment on a daily basis should be continued for at least four weeks. Most horses tolerate an indwelling catheter very well. Surgical occlusion of the artery may cause spontaneous regression of the lesion without antifungal treatment.

The major part of the epithelial lining of the guttural pouch mucosa consists of ciliated cells. Systemic treatment with a β_2-mimetic agent enhances ciliary activity, especially in the first few weeks of treatment when there is still substantial damage to the mucosa. Anti-inflammatory treatment may be necessary during the first few days of treatment. The use of non-steroidal anti-inflammatory drugs is preferred over corticosteroids because of the immunosuppressive effects of the latter. To facilitate guttural pouch clearance, the owner should be instructed to present food from a low level.

236 i. Occult sarcoid.
ii. Surgical intervention can activate the tumour and make it more difficult to treat. Relapse rates after conventional surgery are over 50%
iii. In the initial trials, intralesional injection of cisplatin in oil was reported to be uniformly effective. 5-Fluorouracil has also been reported to be effective in small sarcoids.

237–239: Questions

237 Neonatal maladjustment syndrome, as shown in **237**, is a common problem where normal foal behaviour is affected. The suckle response is usually lost, and foals require extensive supportive care. Give at least two answers to each of the following questions:
i. What are some other behavioural synonyms for neonatal maladjustment syndrome in foals?
ii. What insults to the brain could be responsible?
iii. What brain lesions may be present?
iv. What medications could be provided?

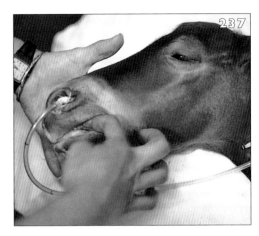

238 Which parts of the coagulation cascade does the assay for activated coagulation time test:?

239 A 12-year-old Thoroughbred brood mare with a one-year history of a loud heart murmur is found dead two days after foaling. Post-mortem evaluation of the heart revealed the depicted abnormality of the mitral valve (**239**).
i. What is the abnormality?
ii. Characterize the type of murmur that might be auscultated in a horse with this problem.
iii. What are possible sequelae to haemodynamically significant regurgitation of the mitral valve?

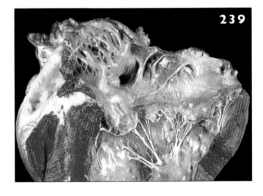

237–239: Answers

237 i. Dummies, barkers, wanderers, sleepers.
ii. Anoxia, hypoxia, sepsis, haemorrhage, meningitis.
iii. Cerebral oedema, haemorrhage, meningitis.
iv. Dimethylsulphoxide, mannitol, oxygen, etc.

238 Intrinsic and common pathways.

239 i. Ruptured chordae tendinae.
ii. A systolic murmur audible over the left side would be present. A musical or 'honking' quality to the murmur may also be heard. A precordial thrill is sometimes palpable.
iii. Severe mitral regurgitation can lead to left atrial and left ventricular dilation, which can lead to atrial fibrillation. In long-standing cases, congestive heart failure will ultimately occur. Clinical findings of congestive heart failure in the horse include atrial fibrillation, tachycardia, jugular pulsations and oedema. In cases of rupture of the chordae tendinae of a major leaflet of the mitral valve, acute pulmonary oedema may occur.

Pulmonary hypertension may become quite severe in horses with left-sided heart failure, and the pulmonary artery may even rupture in some cases. In this mare the left atrium ruptured, leading to a haemopericardium and cardiac tamponade.

Abbreviations

ABC, airway, breathing, circulation
ACD, acid citrate dextrose
ACTH, adrenocorticotropic hormone
AHS, African horse sickness
AI, artificial insemination
ALD, arrested larval development
APTT, activated partial thromboplastin time
AST, aspartate transaminase
BZ, benzimidazole
CBC, complete blood count
CK–MB, creatine kinase–muscle brain
CL, corpus luteum
CID, combined immunodeficiency syndrome
CNS, central nervous system
COPD, chronic obstructive pulmonary disease
CSF, cerebrospinal fluid
DDSP, dorsal displacement of the soft palate
DIC, disseminated intravascular coagulation
DSS, dioctyl sodium succinate
DTM, dermatophyte test medium
eCG, equine chorionic gonadotropin
ECG, electrocardiogram
EDTA, ethylenediaminetetraacetic acid
EHV, equine herpesvirus
EIA, equine infectious anaemia
EIPH, exercise-induced pulmonary haemorrhage
ELISA, enzyme linked immunosorbent assay
EMG, electomyograph
EPM, equine protozoal myeloencephalitis
EVA, equine viral arteritis
FECRT, faecal egg count reduction test

FWEC, faecal worm egg count
GGT, γ-glutamyltransferase
GI, gastrointestinal
hCG, human chorionic gonadotropin
HR, heart rate
IMPT, immune-mediated thrombocytopenia
MCV, mean corpuscular volume
MRI, magnetic resonance imaging
Nd:YAG, neodymium:yttrium aluminium garnet
NSAID, non-steroidal anti-inflammatory drug
NSE, nephrosplenic entrapment
PCV, packed cell volume
PEH, progressive ethmoidal haematoma
PGF-2α, prostaglandin F-2α
PMN, polymorphonuclear neutrophil leucocytes
PMSG, pregnant mare serum gonadtropin
POMC, pro-opiolipomelanocortin
PT, prothombrin time
RBC, red blood cell
RDPPA, rostral displacement of the palatopharyngeal arch
RID, radial immunodiffusion
RR, respiratory rate
SDF, synchronous diaphragmatic flutter
SPAOPD, summer pasture associated obstructive pulmonary disease
TPN, total parenteral nutrition
TRF, thyrotropin releasing factor
TRH, thyrotropin releasing hormone
TSH, thyroid stimulating hormone
VPD, ventricular premature depolarization
VSD, ventricular septal defect
WBC, white blood cell

Index

Question and answer numbers are given in the index, not page numbers.

ABC (airway, breathing, circulation), 132
abdominal fat necrosis, retroperitoneal, 217
abdominocentesis, 13, 120, 128, 144, 165, 196, 227
abortion, 4, 7, 23, 160, 181
acidosis, 90, 106, 211
ACTH, 65, 88
Actinobacillus equuli, 159
adrenalectomy, 65, 134
African horse sickness, 85
alfalfa, 51
alimentary lymphosarcoma, 105, 214
'alkali disease', 9
alkalosis, 45, 106
alopecia, 163, 164, 190
alsike clover, 98, 117
anaemia,
  equine infectious, 66, 123, 171, 186
  haemolytic, 66
anoestrus, 12
Anoplocephala perfoliata, 129, 206
anorexia, 193
anthelmintic resistance, 16, 27
anti-arrhythmic therapy, 208
arthritis, septic, 159
artificial insemination, 4, 48, 57, 67
arytenoid chondritis, 10
ascarids, 124, 150
aspartate aminotransferase, 6, 9
Aspergillus spp., 41, 92, 146
atrial fibrillation, 75, 102, 205, 226, 239
atrioventricular block, 107
azoturia, 204

babesiosis, 66
Bacillus (Clostridium) pilliformis, 84
bicarbonate treatment, 45
bilirubin, 169
biopsy, 6, 99, 125, 163, 166, 169, 192
bladder, sand-like mass in, 231
bladder atony, 101
'blind staggers', 9
blindness, 157, 176
blood transfusion, 32, 66
bone maturity, 60
borreliosis, 175
botryomycosis, 93
botulism, 1, 172
bowel torsion, 87

bronchiolitis, chronic, 113
bronchodilators, 179
caecocaecal intussusception, 189, 223
caecocolic intussusception, 189, 223
calcific keratopathy, 56
calcium, intravenous, 45
calculi, 37
Campylobacter, 125
carotid artery ligation, 235
cataract, 119, 230
catecholamines, 134
catheter, indwelling, 235
cauda equina, neuritis of, 110
cellulitis, 3
cerebral oedema, 237
cervical compressive myelopathy, 215
check ligament desmotomy, 178
chest drain, 72
cholangiohepatitis, 169
cholesterol granuloma, 123
cholesterolaemic granuloma, 58, 123
chordae tendinae, ruptured, 53, 239
chorioptic mange, 71
choroid plexi, 58
chromosomal abnormality, 42
chronic obstructive pulmonary disease, 39, 113, 179
cleft palate, 86
Clostridium spp., 2, 66, 84, 152, 165, 180
Clostridium botulinum, 172
Clostridium tetani, 91, 229
clotting, 34, 126, 182
coagulation cascade, 238
Coggins test, 105
colic, 22, 64, 76, 87, 94, 98, 118, 129, 133, 148, 150, 162, 169, 189, 106, 107, 206, 214, 227, 234
  foal, 13, 20, 21, 29
colitis, 152
  uraemic, 87, 180
colon displacement, 52
colon inflammation, 96
colon obstruction, 76
colostrum, 159, 201, 220
combined immunodeficiency syndrome, 130
congenital abnormality, 29
congestive heart failure, 226, 239
corneal–conjunctival squamous cell carcinoma, 147

Index

corneal infection, *Pseudomanas* spp., 209
corneal stromal abscess, 221
corneal ulcer, 19, 110
corticoid:creatinine ratio, 88
cortisol, 65, 88
creatine kinase, 6, 9
Crohn's disease, 125
cryptosporidiosis, 152, 197
CSF discolouration, 216
CT scan, 157
Culicoides spp., 85
Cushing's disease, equine, 65, 88, 233
cyathostome infection, 16, 27, 152, 173, 192, 223
cystic glandular distension, 166

dehydration, 81, 94, 133
dental tartar, 11
dermatophilosis, 210
Dermatophilus congolensis, 130
dermatophytosis, 69, 190
dermoid lesion, congenital, 213
dexamethasone suppression test, 88, 233
diabetes mellitus, 65, 233
diarrhoea, 2, 16, 152, 173
 foal, 50, 152, 197
Digitalis purpurea, 184
digoxin intoxication, 184
dioestrus, prolonged, 12
disseminated intravascular coagulopathy, 151
dorsal wall stripping/resection, 30
drug reaction, 41
duodenitis, 165
dysmaturity, 198
dysphagia, 92, 189

echocardiograms, 75, 102, 107, 132, 174, 205, 208, 226
Ehrlichia risticii, 127, 187
Eimeria leuckartii, 50
electroencephalogram, 154, 157
ELISA test, 220
embryo crushing, 67
encephalomyelitis, 123
endocarditis, bacterial, 203
endometrial cups, 181
endometritis, 166, 181
endoscopy, 1, 143
endotoxaemia, 34, 81, 94, 128, 148, 152
enteritis,
 anterior, 162, 165
 granulomatous, 105, 125
Enterobacteria sp., 169
enterocolitis, necrotizing, 152
enteroliths, 51, 76

enterotomy, 76
eosinophilic dermatitis, 210
epiglottic entrapment, 62
epistaxis, 151, 219, 235
equine chorionic gonadotrophin, 181
equine coital exanthema, 188
equine Cushing's disease, *see* Cushing's disease
equine degenerative myeloencephalopathy, 215
equine dysautonomia, *see* grass sickness
equine exfoliative eosinophilic dermatitis, 210
equine herpesvirus, 23, 61, 149, 188, 212
equine infectious anaemia, *see* anaemia, equine infectious
equine influenza, 61, 224
equine monocytic ehrlichiosis, *see* Potomac horse fever
equine motor neurone disease, 17
equine protozoal myeloencephalitis, 89, 185, 215
equine sarcoidosis, 210
equine viral arteritis, 23, 80
erythema multiforme, 114
erythrocytes, 9
Escherichia coli, 152, 159
ethmoidal haematoma, progressive, 126
euthyroid sick syndrome, 164
eyelid laceration, 74

facial paresis/paralysis, 55, 89, 110
factor VII consumption, 34
faecal egg counts, 27
fatty acids, 211
femur, fractured, 217
fertility, 156
fetal reduction, 4
fetal membrane bacteria/fungi, 7
fibrin, 118, 151
fibrosis,
 hepatic, 117
 periglandular, 166
'flail' chest, 191
flexor tenotomy, 178
fluid therapy, 94, 148
forelimb contracture, 164
frog support, 46
furunculosis, infectious, 139

gastric capacity, 78, 162
gastric rupture, 18
gastric ulceration, 21, 207
Gastrophilus intestinalis, 109
gastroscopy, 207
gestation, length of, 198

189

Index

GGT elevation, 98
glanders, 168
glomerulonephritis, 49, 99
glucose tolerance test, 214
goitre, 164
gonadal dysgenesis, 63X, 28, 42
grain overload, 148
granulomatous enteritis, 105, 125
granulosa cell tumour, 100, 143, 156
grass sickness, 83, 133, 138, 141, 162, 189,
guttural pouch, 54, 92, 126, 146, 200, 235

haematoma, 70, 121, 126, 143
haemolysis, immune-mediated, 66
haemolytic disease, 193
haemoperitoneum, 191
haemosiderin, 8
haemothorax, 191
heart failure, congestive, 226, 239
Heinz bodies, 14, 15
Helicobacter pylori, 21
hepatic failure, 193
hepatic fibrosis, 117
hepatitis, serum, 225
hepato-encephalopathy, 34
hepatopathy, steroid, 65
herniorrhaphy, inguinal, 95
hirsutism, 39, 65, 233
hocks, 'curbed', 60
hormones, 88, 112, 143, 164, 194
Horner's syndrome, 146
Howell–Jolly bodies, 15
human chorionic gonadotrophin, 48
hydrocephalus, 154, 157
hypercalcaemia, 194, 214
hyperkalaemia, 90
hyperlipaemia, 59, 184, 211
hyperthyroidism, 47
hypertrophic osteopathy, 103
hypervitaminosis D, 194, 214
hypocalcaemia, 22, 45, 64
hypochloraemia, 64
hypothyroidism, 112, 164
hypovolaemic shock, 79

icterus, associated with anorexia, 193
ileocolonic aganglionosis, 137
ileum impaction, 20
immunity, passive transference of, 201, 220
immunodeficiency syndrome, combined, 130
infertility, 28
inguinal herniorrhaphy, 95
insulin, 88, 211
intestinal malabsorption, 105
intestinal obstruction, 29, 162, 206

iris cysts, 167
isoerythrolysis, neonatal, 66, 79
ivermectin, 223

jejunitis, proximal, 165
joint ill, 159
jugular distension, 26

karyotyping, 42
keratitis, 41, 110
kernicterus, 79
kidney, 99, 120, 153, 161

lactation tetany, 22, 45
laminitis, 30, 35, 46, 88, 165, 233
laparotomy, 77, 83, 118, 124, 133, 223
laryngeal hemiplegia, 5
laryngeal neuropathy, recurrent, 5
laxatives, 78
lethal white syndrome, 29, 137
lipoproteins, 211
liver disease, 117, 169, 225
lumbosacral spinal tap, 110, 149
lung field, 222
Lyme disease, 175
lymphosarcoma, 26, 38, 105, 122, 135, 214

magnetic resonance imaging, 154, 157
maladjustment syndrome, 154, 237
Marie's disease, 103
masseter myopathy, 111
meconium impaction, 29
mediastinal lymphosarcoma, 26, 135
megaoesophagus, 86
microhaematocrit tube, 225
milk replacers, 44
Miniature horse, 68, 76, 184
mitral valve disease, 203
mitral valve regurgitation, 75
moxidectin, 223
mucokinetic agents, 179
muscle relaxants, 178
muscular dystrophy, 9
myocarditis, 158, 208, 224
myoglobinuria, paralytic, 204
myositis, 3, 204
myotonia, 155

nasal polyps, 126
nasogastric reflux, 162
nasogastric tube, 44, 78, 165, 234
neoplasia, 123, 126, 135, 153, 164, 212
nephrosplenic entrapment, 52, 77, 162
nidi, 51

Index

Nissl substance, 133
nodular panniculitis, 139
NSAIDs, 37, 96, 234, 235

Obel grades, 35
obstructive pulmonary disease,
 chronic, 39, 113, 179
 summer pasture-associated, 113
oesophageal obstruction, 59, 86, 133, 145
oestrous cycle, 143, 166, 181
oestrus, persistent, 100
oestrus behaviour, 12, 42, 100
omentum tissue, 196
onchocerciasis, cutaneous, 69
onion poisoning, 14, 66
osteodystrophic line, 202
ovarian atrophy, 28
ovarian granulosa cell tumour, 156
ovariectomy, 100, 156, 186
ovary, enlarged, 143
ovulation, two-follicle, 48, 57
Oxyuris, 36

palatopharyngeal arch, rostral displacement of, 228
parafollicular (C-cell) tumour, 47
paraphimosis, 70
Parascaris equorum, 124, 150
parasite prophylaxis programmes, 16
parathyroid glands, 22
passive immunity, 201, 220
PCV measurement, 81
pectoral oedema, 26
pelvic flexure impaction, 78, 234
pemphigus foliaceus, 190, 210
penicillin reaction, 25
penile paralysis/prolapse, 70
pericardial fluid, 82
pericarditis, 26
peripheral vestibular syndrome, 55
peritonitis, 144, 162
phaeochromocytoma, 134
pharyngeal lymphoid hyperplasia, 115
pharyngitis, 115
pituitary abscess, 65
pituitary adenoma, 39
 pars intermedia, 65, 88, 233
pituitary-dependent hyperadrenocorticism, 65
placental lesion, 7
plasma cortisol, 88
plasma transfusion, 220
platelet count, 151, 219
pleural effusion, 38
pleural fluid, 140, 153, 222
pleural lavage, 72

pleurisy, 72
pleurocentesis, 140
pleuropneumonia, 195
pleximeter, 222
pneumonia,
 aspiration, 44, 172
 granulomatous, 212
 haematogenous bacterial, 212
 mycotic, 212
Pneumocystis carinii, 130
portosystemic shunt, 157
post-castration infection, 70
Potomac horse fever, 127, 152, 180, 187
pregnancy diagnosis, 166, 181
prematurity, 60, 198, 201
procaine toxicity, 25
progressive ethmoidal haematoma, 126
prothrombin time, 34
pruritis, 36
Pseudomonas spp., 168, 209
pseudothrombocytopenia, 219
psoroptic mange, 36
pulmonary artery rupture, 203
pulmonary haemorrhage, exercise-induced, 8, 126, 170
pulmonary neoplasia, 153
pulmonary oedema, 25
purpura haemorrhagica, 3, 31
pyrrolizidine alkaloid toxicity, 117, 162

quinidine sulphate, 205, 226

rachitic rosary, 202
radial immunodiffusion test, 220
radiography, 13
radiotelemetry, 107
ragwort, 189
RBC lysis, 79
rectal prolapse, 40
rectal tear, 104
red maple toxicosis, 14, 66
renal adenocarcinoma, metastatic, 153
renal carcinoma, 120
renal crest necrosis, 37
renal disease, 99, 194
renal failure, 11, 37, 49, 90, 199
renal pyramids, 161
retinal detachment, 230
rhabdomyolysis, exertional, 204
Rhodococcus equi, 131, 159
rib fracture, neonatal foal, 43, 191
'roar', 5
rolling, 77
rotavirus infection, 50, 152

Index

sabulous urolithiasis, 231
Salmonella spp., 159, 165, 180, 197
salmonellosis, 127, 152, 180
sand, 128
sarcoid,
  granulation, 93
  occult, 236
sarcoidosis, equine, 210
Sarcocystis neurona, 185, 215
scrotal hernia, 95
seizures, persistent, 154
selenium, 9, 111
septicaemia, 116, 132
serological tests, 61
serum IgG values, 220
serum protein electrophoresis, 192
shoe, heart-bar, 46
sinonasal adenocarcinoma, 219
skin sloughing, 3
soft palate, 33, 62, 86
spleen, 52, 136
squamous cell carcinoma, 109, 147
staphylococcal folliculitis, 190
Staphylococcus aureus, 55, 73, 159
steatitis, 9
stone removal, 142
strangles, 3, 31, 54, 63, 218
stranguria, 101
Streptococcus spp., 24, 159
Streptococcus equi, 3, 54, 63, 176
strongyles, 150, 206, 223
Strongyloides westeri, 152
synchronous diaphragmatic flutter, 22, 64
systolic cardiac murmur, 75

tail rubbing, 36
tendons,
  contracted, 178
  ruptured, 164
tetanus, 91, 225, 229
Theiler's disease, 225
thoracic auscultation, 179
thoracic neoplasia, 135
thoracic percussion, 222
thoracocentesis, 72, 135, 140
thoracotomy, 72
thrombocytopenia, immune-mediated, 186, 219
thromboplastin time, activated partial, 34
thrombosis, 87, 151
thyroid carcinoma, 47
thyroid gland, asymmetrical hyperplasia of, 47
thyroid hormones, 88, 112, 164
thyroid neoplasia, 164

ticks, 175
tongue atrophy, 185
tracheal collapse, congenital, 68
tracheobronchial lymphadenopathy, 131
tracheostomy, 151
trans-sphenoidal pituitary microsurgery, 65
transvaginal ultrasound-guided needle aspiration, 4
tricuspid insufficiency, 226
twins, 4, 57, 67
Tyzzer's disease, 84

ultrasound scans, 13, 26, 38, 57, 67, 120, 166, 174, 182
umbilical artery clot, 182
umbilical cord, twisted, 160
umbilical hernia, 108
urachal abscess, 24
urachal diverticulum, 101
uraemic colitis, 87, 180
urea nitrogen, 211
ureter, ectopic, 177
urethral disorder, 183
urinary calculi, 142
urinary incontinence, 97, 177, 231
urination, bleeding at end of, 183
urine, red, 14
uroperitoneum, 97
urticaria, 114
uterine artery rupture, 121
uterine cysts, 57
uterine tumour, 143
uveitis, 56, 119

vaginal haematoma, 121
vasculitis, immune-mediated, 3
venom-induced inflammatory reaction, 151
ventricular premature depolarization, 158, 208
ventricular septal defect, 174, 226
ventricular tachycardia, paroxysmal, 102
vertebral fracture, 216
vices, 36
villous atrophy, 125
vinegar, 51
virus isolation, 61
vitamin D, 194, 202
vitamin E, 9, 17
vomiting, 18, 162
vulvoplasty, Caslick's, 121

'whistle', 5

zinc sulphate turbidity test, 220

PALO ALTO COLLEGE LRC
1400 W VILLARET BLVD
SAN ANTONIO, TX 78244

DATE DUE

GAYLORD   PRINTED IN U.S.A.

WITHDRAWN

PALO ALTO COLLEGE LRC

Self Assessment colo

36171001423620